Computing Possible Futures

COMPUTING POSSIBLE FUTURES

Model-Based Explorations of "What if?"

WILLIAM B. ROUSE

OXFORD
UNIVERSITY PRESS

OXFORD
UNIVERSITY PRESS

Great Clarendon Street, Oxford, OX2 6DP,
United Kingdom

Oxford University Press is a department of the University of Oxford.
It furthers the University's objective of excellence in research, scholarship,
and education by publishing worldwide. Oxford is a registered trade mark of
Oxford University Press in the UK and in certain other countries

First Edition published in 2019

Impression: 1

Published in the United States of America by Oxford University Press
198 Madison Avenue, New York, NY 10016, United States of America

British Library Cataloguing in Publication Data
Data available

Library of Congress Control Number: 2019939954

ISBN 978–0–19–884642–0

DOI 10.1093/oso/9780198846420.001.0001

Printed and bound by
CPI Group (UK) Ltd, Croydon, CR0 4YY

PREFACE

I have worked with thousands of executives and senior managers during the nonacademic elements of my career, in over 100 enterprises. For the past decade, I have worked with executives of well over 10 large U.S. healthcare providers. They are very interested in what "data analytics" can do for them and, quite recently, what the prospects are for "AI and machine learning."

These executives were trained in science in medical school, where they were familiar with basic math, for example, algebra and statistics. But that was a long time ago. Nevertheless, they really want to understand these topics, their implications, and how they can best invest to take advantage of these trends. As a result, I spend more time explaining modeling to my consulting clients than to my students.

I have had similar experiences with executives in the automotive, aerospace, consumer, electronics, pharmaceutical, publishing, and semiconductor industries. They are well educated and highly motivated. They want to understand and then invest wisely. And—they are reasonably skeptical. They have often experienced overselling and under-delivery.

They ask about reasonable and realistic expectations. *Computing Possible Futures* addresses this question. Their concern is with the *futurity* of decisions they are currently entertaining. They cannot fully address this concern empirically. The future does not yet exist to be measured. Thus, they need some way to make predictions.

Various pundits, and perhaps a few oracles, will confidently tell them what to expect. Most executives want more rigor than this. Computational modeling can usually provide substantially more rigor than expert, or not so expert, opinion can. These models can be used to predict the future.

Executives want these predictions to be accurate. The problem is that we rarely can predict exactly what will happen, only what *might* happen. To overcome this limitation, executives can be provided predictions of possible futures and the conditions under which each scenario is likely to emerge. Models can help them to understand these *possible* futures.

I have found that most executives find such candor refreshing, perhaps even liberating. Their job becomes one of imagining and designing a portfolio of possible futures, assisted by interactive computational models. Understanding and managing uncertainty is central to their job. Indeed, doing this better than competitors is a hallmark of success.

Computing Possible Futures is intended to help them to understand what fundamentally needs to be done, why it needs to be done, and how to do it. Such readers are unlikely to be the ones who will create the computational models. These executives will recruit managers who will hire and guide the modelers. My hope is that all of these people will read and discuss *Computing Possible Futures*, developing a "shared mental model" in the process, which greatly enhances their chances of success.

Who else are intended readers of *Computing Possible Futures*? Those who want to sell modeling engagements to executives will find that this book helps them. Accenture, Bain, Boston Consulting Group, IBM, and McKinsey are among this group, collectively employing hundreds of thousands of consultants. *Computing Possible Futures* is intended to provide the "lingua franca" among the consumers and providers of computational modeling.

Computing Possible Futures is not a textbook. There is not enough theory and math for most faculty members. Students will, however, want to read this book to imagine working for companies that employ this approach to understanding and managing the complexities of their markets. They may want to imagine working for the above consulting companies. These students might be pursuing MBAs, but the engineer in me leads to my belief that this book will be more popular with aspiring and recently minted engineers.

Much of the material in *Computing Possible Futures* is case based, with stories of how I have used the models to support executives and senior managers to make well-informed decisions. Many people have told me that these stories are what bring model-based decision-making to life. My sense is that interesting and insightful stories are the best way to communicate and provide readers motivations to dig deeper.

Scores of people have influenced my thinking in this arena and contributed to the stories that I relate. With over 50 years of experiences, acknowledging everyone who has played a part would consume too much space. I would easily be at risk of forgetting people who played important roles. Thus, I will keep this simple. Thank you.

William B. Rouse
Washington, DC
January 2019

CONTENTS

CHAPTER 1

Introduction

There are many "what is?" questions. What is the temperature outside? How many people are in line at customer service? What is the shortest route from A to B? How many people graduated from high school last year? With the right data, including your own observations, these questions can readily be answered.

"What if?" questions are different. They cannot be answered empirically because the future you are considering does not yet exist. How long would it take if we walked instead of drove? How many people will graduate over the next decade? What if we moved our investments from X to Y?

Addressing these types of question requires predictions. Sometimes these predictions come from our mental models of the phenomena of interest. For example, we may know the streets of the city pretty well and have experienced past traffic patterns. We use this knowledge to imagine the time required.

We might extrapolate to answer the question. We know the number of children alive at the right ages to graduate over the next decade. We can access actuarial tables on mortality to predict how many will make it to high school. We don't know what will happen economically or socially over the next decade, so we expect the predictions we calculate will have some associated uncertainty.

The notion of our predictions being uncertain is central to the discussions in this book. We cannot know what *will* happen, but we can predict what *might* happen. Further, a range of futures might happen. This range will include *possible futures*. Many futures will be very unlikely. That is useful to know as well.

How can you project the possible futures relative to the "what if?" question of interest? Beyond using your mental models and your imagination, you can

Computing Possible Futures. William B. Rouse. Oxford University Press (2019). © William B. Rouse 2019.
DOI: 10.1093/oso/9780198846420.001.0001

employ computational models. These exist in a variety of forms, ranging from quite simple to very sophisticated.

The simplest form is proverbial "back of the envelope" estimates, which, in my experience, are more often on tablecloths or cocktail napkins than envelopes. I often try to estimate the answers I expect from computational models. Using simplifications that make calculations easier, I derive a very rough estimate of what I expect. If my estimate differs significantly from what the computational model provides, I track down the source(s) of the discrepancy.

From this traditional first step, you might next formulate a set of mathematical equations. This usually starts with paper and pen and perhaps results in a spreadsheet model. This can provide more precise predictions and, hopefully, more accurate predictions. Of course, this by no means eliminates the uncertainties noted above.

The next formulation may transition the spreadsheet model to a computer program that enables including uncertainties and provides a variety of visual representations of predictions. This may involve using one or more commercially available software tools, several of which are noted in later chapters.

Simple Examples

This section uses some simple examples to illustrate the issues raised above. My goal is to set the stage for less simple examples later in this chapter, as well as much more elaborate stories in later chapters.

Waiting in Line

Figure 1.1 portrays a classic waiting line. Customers are concerned with how long they will have to wait to be serviced. We want to predict this for them.

We can use queuing theory to model this waiting line. Assume that customers arrive randomly at an average rate of λ customers per hour and are serviced at an average rate of μ per hour. The utilization of this queuing system, ρ, is defined as the ratio λ/μ. The mean number of people in the system, L, including the person being served, equals $\rho/(1 - \rho)$. The standard deviation is the square root of $\rho/(1 - \rho)^2$.

It is clear that $\rho < 1$. Otherwise, L becomes infinite, because the server can never catch up with the number of people waiting. If $\rho = 0.9$, then L equals 9, and the standard deviation of the number of people waiting is, roughly, 9 as

Fig 1.1 Waiting in line.

well. The answer to the customer's question is that there will, on average, be 8–9 people ahead of them, but sometimes it will be many more. With $\rho = 0.9$, there will often be significant waiting time. So, the possible futures for customers are quite varied.

This is a standard and quite simple queuing model. A few key aspects of the real problem are missing. Customers, upon seeing the waiting line, may "balk" and not enter the line. Other customers may wait a bit but then "renege" and leave the waiting line. Both of those behaviors will benefit those who remain in line, but they may not help the service organization's reputation. To avoid this, the organization may add a second server, which would require a significant expansion of this model.

In this example, we have portrayed the range of possible futures in terms of a mean and a standard deviation. Any "point" prediction, for example, predicting exactly nine people in the system, would almost always be wrong. This is due to random variations of interarrival and service times, as well as the simplicity of the model, for example, no balking or reneging. Nevertheless, such a model could be a reasonable first step. It at least tells that we need more than one server, or a spacious waiting room.

Investing for Retirement

This is a problem almost everybody eventually faces. You need money for your living expenses, which are denoted by LE. Due to inflation, at an annual rate of

α, this amount increases with years. In anticipation of retirement, every year you invest money, which is denoted by RI, and earn investment returns at an annual rate of β. Over time, these retirement assets, denoted RA, steadily grow.

Assume you do this for 20 years and then retire. At this point, you draw from RA to pay for LE, no longer investing RI each year. What should RI equal during the years that you are still working and investing?

Figure 1.2 portrays the results when α is 3 %, β is 7 %, LE equals \$100,000, and RI equals \$50,000, assuming that the latter amount is held constant over the working years. Exponential growth of your assets during the first 20 years is replaced by accelerating decline of, starting in Year 21. The acceleration is due to the fact that your living expenses are steadily increasing because of the 3 % inflation, from \$100,000 in Year 1 to \$317,000 in Year 40. Despite the ongoing 7 % return on your remaining assets, your living expenses drain these assets. In contrast, if inflation were 0 %, your assets would steadily increase over the 40 years, leaving your heirs with a tidy inheritance. Alternatively, you could decrease your annual investment to \$26,000 to break even in Year 40.

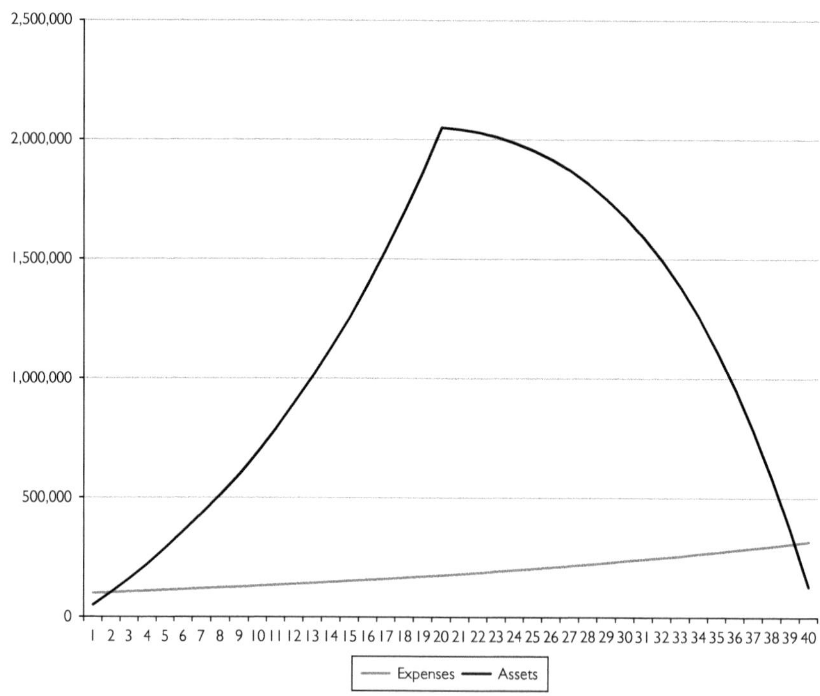

Fig 1.2 Expenses and assets over 40 years.

This seems like a fairly straightforward model, not unlike those embedded in many retirement planning applications on the Internet. However, it only provides point predictions for a process with substantial volatility. Inflation rates and investment return rates are highly variable.

Thus, the projections shown in Figure 1.2 look much more certain than they are in reality. Of course, one can update such projections each year and, if fortunate, adjust investments for any surprises. However, starting in Year 21, you have no more working income to use to compensate for these surprises. This seems to have caused many people to put off retirement.

This model is useful in that it illustrates the enormous impact of even modest inflation compounded over 40 years. Yet, it does not portray the different scenarios that might play out. We could vary α and β to determine the RI needed for each combination. This would constitute an elementary form of "scenario planning," a topic addressed in Chapter 3.

Driving a Vehicle

We would like to predict the performance of a vehicle, for example, time to accelerate from 0 to 60 miles per hour, and stability on curves, as a function of the design of the vehicle. This requires that we predict driver performance in the manual control of the vehicle. This prediction problem is addressed in depth in Chapter 6, but is simplified here.

The block diagram in Figure 1.3 represents the driver–vehicle system. Using a model based on this figure, we can predict the performance of this system. These predictions would likely be reasonably accurate for drivers maintaining lane position and speed on straight roads in clear weather in daylight. However, reality includes irregular roads, wind, rain—and other drivers!

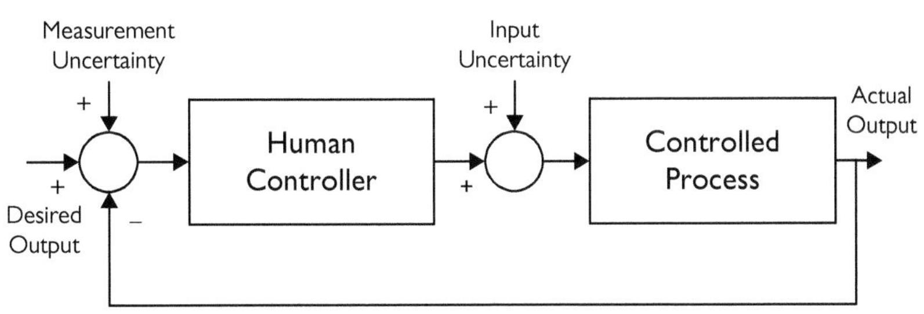

Fig 1.3 Driver–vehicle system.

This model includes a desired output—the intended path—measurement uncertainty representing possible limited visibility, and input uncertainty representing driver variability in steering, accelerating, and braking. This model should yield reasonable predictions, but it does not include important factors like traffic flow, other drivers' behaviors, and possible distractions, for example, fussing with the entertainment system. Many sources of uncertainty and variability are missing. Nevertheless, this model would be useful for predicting driver–vehicle performance in, admittedly, pristine conditions.

This is a good example of a model that might be used to diagnose poor designs but is not sufficient to fully validate good designs. It is common for such validation to happen in human-in-the-loop driving simulators that combine computational models of the vehicle with live human drivers in simulated realistic driving conditions.

Less Simple Examples

At this point, I move from hypothetical examples, which were created to illustrate key points, to real models developed to help decision-makers answer questions of importance. Nevertheless, the depth with which these two examples are discussed is limited to that necessary to further elaborate the issues raised in this chapter.

Provisioning Submarines with Spare Parts

My first engineering assignment when I worked at Raytheon was to determine the mix of spare parts that a submarine should carry to maximize the system availability of the sonar system. Availability is the probability that a system will perform as required during a mission, which, for a submarine, could be months.

System availability is affected by failure rates of component parts, repair times for failures, and relationships among components. Functional dependencies and redundancies affect the extent to which a failed part disables the overall system. Redundant parts provide backups to failures. Figure 1.4 depicts a notional sonar system.

The simulation model developed was called MOSES, for Mission Oriented System Effectiveness Synthesis (Rouse, 1969). MOSES included a representation of the system similar to Figure 1.4. Data for the thousands of parts in the

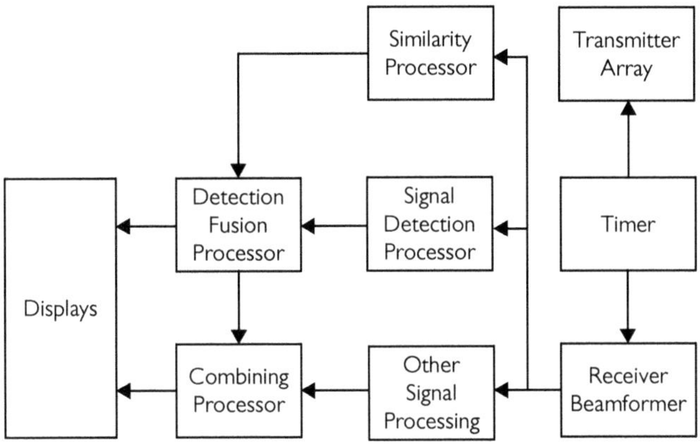

Fig 1.4 Notional sonar system.

system, obtained from contractually required testing, included the mean time between failures (MTBF) and the mean time to repair (MTTR) for each part.

MOSES simulated operation of the sonar system over long missions such as, for example, a 30-day mission. As failures occurred, the structure of the system (Figure 1.4) was used to determine the impact on system operations, that is, whether or not the system was available. Maintenance actions were also simulated, solely in terms of time required to replace the failed part. Across the whole simulation of the mission and system operations, the overall availability statistic equaled MTBF/(MTBF + MTTR).

Typically, the MTBF is much larger than the MTTR. Otherwise, availability suffers. On the other hand, if repairs are instantaneous, that is, MTTR equals zero, availability is 100 %, regardless of the MTBF. This might seem extreme, but redundant parts that are automatically switched into use are increasingly common.

Determining the best mix of spare parts involved finding the mix that maximized availability within the physical space constraints of the submarine. MOSES was also used to determine where increased redundancy would most improve availability. The number of requests for various MOSES analyses resulted in many long simulation runs, which led the personnel in the computer center to comment that MOSES was often wandering in wilderness.

This example involved a complicated system of which we had ample knowledge because all of it was designed or engineered. There were no humans interacting with the system. MOSES did not consider whether the targets and threats detected by the sonar system were successfully engaged. In other words, the predictions were of availability rather than effectiveness. In contrast, for many of the examples in this book, behavioral and social phenomena are central, which significantly complicates modeling endeavors.

Thus, the possible futures computed by MOSES were limited to predictions of the percentage of times the sonar system would be available for use. This percentage was strongly affected by system redundancies and the mix of spare parts available. The simulation also yielded an estimated probability distribution of this percentage.

This distribution provided a means to estimate the risks associated with the predictions. Contractual requirements dictated that mean availability exceed a specified level. Using the probability distribution, we could estimate the risk that the actual availability experienced on a mission would be less than this level despite the mean exceeding it. We knew we needed more than just a point prediction.

Routing Requests for Information Network Services

The State of Illinois supports the Illinois Library and Information Network (ILLINET). The purpose of this network is to provide information services to the citizens of the state—all those with library cards or equivalent. Services range from answering reference questions to providing interlibrary loans.

Service requests are routed through a multilevel network that includes local, regional, and state resources—3,000 organizations in all. There are 18 regional sub-networks and four major statewide resources. The model of each regional sub-network is show in Figure 1.5. Requests for services enter Node 1. Successes depart from Node 4, and failures from Node 6. Failures are routed to other resources for potential fulfillment. Delays and possible service failures are, therefore, encountered at each node.

We developed a queuing network model of ILLINET (Rouse, 1976). This interactive model was used to determine the optimal path through the multilevel network for each type of service. Overall measures of performance included

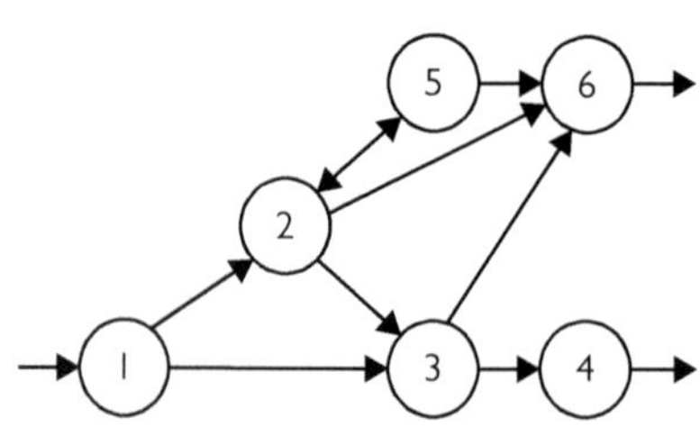

Fig 1.5 Processing at each regional service organization.

probability of success (P), average time until success (W), and average cost of success (C). Not surprisingly, overall performance is strongly affected by aggregate demands across all services, that is, the lengths of waiting lines for services.

The model developed was purely analytical rather than a simulation. It involved quite a bit of matrix algebra, where the matrices reflected network and sub-network structures as well as routing vectors. Parameters within the model included arrival rates for each class of request and service rates for each node in the multilevel network. Data to estimate these parameters came from many thousands of past service requests. Multilevel enterprise models are discussed in depth in Chapter 7.

Routes were determined to maximize an overall utility function $U(P, W, C)$ with component utility functions $U_P(P)$, $U_W(W)$, and $U_C(C)$. The relative weights on these component utility functions could be varied, depending on decision-makers' preferences. Utility theory is discussed in detail in Chapters 2 and 4.

The ILLINET model was used in a variety of ways over several years; some of these uses surprised us, as we had not anticipated these modes of use. The most memorable application provides a compelling example that illustrates one of the themes of this book.

We determined the optimal routing vectors across all types of requests. We found that service could be maintained at current levels for 10 % less budget. Alternatively, service (P and W) could be improved without increasing the budget. The key was to avoid one of the largest statewide resource centers, which was very inefficient and expensive. Making this resource center a "last resort" saved a large portion of the 10 % but also greatly diminished this resource center's revenue.

A meeting was held in Springfield, the state capitol, to discuss this policy. The director of ILLINET and the director of the problematic resource center participated, along with other staff members and our team. The ILLINET director asked the resource center director,

> "Do you see what you are costing me?"
> "Yes, I understand but there is not much I can do about it," he responded.
> "I know and I am willing to ignore this result, if you will relent on the other issue we have been arguing about for the past several months," she offered.
> "Deal. I can do that."

When the meeting ended, I remained to chat with the ILLINET director.

> "What did you think of the model?" I asked.
> "Wonderful. It gave me the leverage to get what I needed," she replied.

This example illustrates the variety of ways in which decision-makers employ models. They will seldom just implement the optimal solution. They are much more oriented to looking for insights, in this case, insights that provide leverage in a broader agenda. Models are a means to such ends rather than ends in themselves.

Key Points

"What is?" is often an important question. However, this book focuses on "what if?" questions, which require predictions to answer. These predictions can seldom specify what *will* happen, so we inevitably address what *might* happen. There are often many possible futures. Hence, single point predictions are often misleading. Instead, we need to explore a range of scenarios, identifying leading indicators and potential tipping points for each scenario.

Beyond being unable to predict precisely what will happen, we can use models to explore designs of systems and policies to determine whether these designs would be effective in admittedly simplified situations. If designs are ineffective in such situations, they are inherently bad ideas relative to the realities in which they are intended to operate.

Models are means to ends rather then ends in themselves. Decision-makers seldom crave models. They want their questions answered in an evidence-based manner. They want insights that provide them competitive leverage. They want to understand possible futures to formulate robust and resilient strategies for addressing these futures.

Overview

The examples discussed thus far provide a sense of the flavor of this book. There are many case studies, numerous diagrams, and almost no equations. My goal is to provide enough detail so that the case studies feel real, without requiring readers to delve into the details of each modeling paradigm.

Chapters 3–9 address the overarching questions shown in Figure 1.6. A wide range of models are discussed in terms of how these models were used to address these questions. Chapter 10 provides a summary of key points throughout the book.

Economies *How Bubbles Happen & Why They Burst*
Markets *How Products & Services Compete*
Technologies *How Options Enable Innovations*
Failures *How to Detect & Diagnose Malfunctions*
Health & Well Being *How to Scale Results to Populations*
Augmentation *How Intelligent Systems Technology Can Help*
Transformation *How Enterprises Address Fundamental Change*

Fig 1.6 Questions of interest.

Chapter 2: Elements of Modeling

This chapter begins by considering definitions of the noun "model." Common definitions include: (1) an exemplar to be followed or imitated and (2) 3D representations of objects, typically on a smaller scale. This book is concerned with representational models, depictions, which reflect the nature of phenomena of interest, for example, structural descriptions, or sets of equations. Also important are computational models, that is, hardware and/or software code that enables the "solution" of a representational model.

Representational models depict phenomena. The five examples discussed earlier were based on a range of depictions of phenomena—queues, cash flows, control systems, networks of components, and networks of flows. These depictions represent abstractions of physical reality. Computational models enable "solving" representational models in the sense of computing the evolution of the phenomena represented, typically over time. These computed futures are predictions of the evolution of phenomena.

This chapter considers the process of modeling in some detail (Rouse, 2015). This process includes framing the questions of interest and identifying the phenomena that will need to be addressed to answer the questions of interest. I discuss eight classes of phenomena that are potentially of relevance.

The next steps are visualization of phenomena, representation of phenomena, and computationally "solving" the representations to predict the impacts of inputs on outputs. The concept of "system state" is discussed. I then outline eight alternative representations: dynamic systems theory, control theory, estimation theory, queuing theory, network theory, decision theory, problem-solving theory, and finance theory.

The exposition in this section may seem rather complicated. However, there are only a few central concepts—state, dynamic response, feedback control, uncertain states, discrete flows, networks flows, decision trade-offs, systems failures, and economic value.

This chapter concludes with a discussion of types of model validation. Two questions that sponsors of modeling efforts often ask are reviewed: first, "How wrong can I be and still have this decision make sense?" and, second, "How bad can things get and still have this decision make sense?" The key point is that sponsors are most concerned with the validity of the decisions they make using models.

Chapter 3: Economic Bubbles

Point predictions are almost always wrong. The next chapter addresses this in the context of economic bubbles, ranging from "tulipomania" in the 17th century, to the recent real estate bubble, to the current cost bubble in higher education. This can be due to the impossibility of getting all of the underlying assumptions precisely correct. Recall, for instance, the retirement planning example discussed earlier in this chapter, where we assumed inflation was constant for 40 years. It would be better to predict probability distributions of outcomes, based in part on probability distributions associated with assumptions.

Another reason that point predictions seldom make sense is the possibility that multiple scenarios are evolving, and it is not clear which one(s) will dominate over time. Scenarios are useful for understanding leading indicators of their emergence and potential tipping points whereby scenarios become inevitable. In some cases, multiple models are used to make such predictions, as is done when predicting courses and impacts of hurricanes. This chapter illustrates this process, using four scenarios for the future of research universities to explore the conditions under which the higher-education cost bubble is likely to burst (Rouse, 2016; Rouse, Lombardi, & Craig, 2018).

Chapter 4: Markets and Competitors

I next move from economies to markets, with particular emphasis on how products and services compete in markets. The discussion is framed in terms of an approach we developed, which we termed "human-centered design" (Rouse, 2007). This approach addresses the concerns, values, and perceptions of all the stakeholders in designing, developing, manufacturing, buying, using, and servicing products and systems.

The *Product Planning Advisor*, a modeling tool for human-centered design, is presented. This tool integrates a construct termed "quality function deployment" with multistakeholder, multiattribute utility theory. Numerous case examples are discussed, ranging from designing the Mini Cooper to developing autonomous aircraft. Two other expert system-based tools are discussed briefly—the *Business Planning Advisor* and the *Situation Assessment Advisor.* These two tools provide examples of technical success but market failure compared to the *Product Planning Advisor.* This leads to a discussion of what users want from model-based tools.

Chapter 5: Technology Adoption

New product and service offerings often require new technologies to be competitive. This, in turn, requires upstream R & D to yield downstream technology innovations. This upstream R & D can be seen as providing "technology options" to the enterprise's business units. The business units can later exercise one or more of these options if it makes competitive sense at the time.

In this chapter, I consider how to attach value to technology options. I address the question "What are they worth?" This can be contrasted with the question "What do they cost?" Worth usually has to greatly exceed cost to garner investments. In general, everyone wants to gain value greater than the purchase price.

The *Technology Investment Advisor* provides a means to answering the worth question. Using option-pricing models and production learning curves, this model projects an enterprise's future financial statements based on the upstream options in which it has invested.

Options-based thinking also provides a framework for value-centered R & D. I discuss ten principles for characterizing, assessing, and managing R & D value. An enterprise simulation, *R & D World*, is used to assess the merits of alternative measures of value. I summarize numerous real-world case studies.

This chapter concludes with an in-depth consideration of technology adoption in automobiles. I discuss several model-based studies, including an assessment of the best ten and worst ten cars of the past 50 years; a study of twelve cars that were withdrawn from the market in the 1930s, 1960s, and 2000s; and studies about the adoption of new powertrain technologies and driverless car technologies.

Chapter 6: System Failures

Modeling efforts are more challenging when behavioral and social phenomena are significant elements of the overall phenomena being addressed. This chapter discusses approaches to modeling human behavior and performance. Human tasks considered include estimation, manual control, multitask decision-making, and problem-solving.

These tasks are addressed in the context of human detection, diagnosis, and compensation for system failures. How do people detect that something has failed, diagnose the cause of the failure, and compensate for the failure to keep the system functioning? Humans are often remarkable at this, but not always. I explore why this is the case.

Detection requires decision-making. Diagnosis involves problem-solving. Compensation focuses on control. Thus, modeling human performance in this arena requires theories of decision-making, problem-solving, and control. The concepts underlying these theories are discussed and illustrated with numerous examples of system failures and how humans addressed them.

I discuss the notion of mental models, including individual versus team mental models and the differences between experts and nonexperts. Behavioral economics is discussed in terms of biases and heuristics, as well as how "nudges" can change behaviors (Kahneman, 2011; Thaler & Sunstein, 2008). Training (which improves humans' potential to perform) or aiding (which directly augments humans' performance) can enhance humans' tasks. I discuss modeling trade-offs between training and aiding.

Chapter 7: Health and Well-Being

This chapter illustrates how computational modeling can substantially contribute to exploring possible futures for health and well-being. I discuss how patient, provider, and payer data sets can be used to parameterize these computational models. I consider how large interactive visualizations can enable a wide range of stakeholders to participate in exploring possible futures.

Policy flight simulators are discussed in terms of how they can enable projecting likely impacts of policies, for example, alternative payment schemes, before they are deployed. These simulators can help to address the enormous variety of participants in healthcare, including patients, providers, and payers, as well as the economic and social circumstances in which they operate. Computational models can be invaluable for projecting the impacts of this variety and considering how system and policy designs should be tailored.

Multilevel enterprise models provide a means for addressing enterprise ecosystems (Rouse, 2015). These models address the people, processes, organizations, and society forces in an enterprise ecosystem. Representative models include macroeconomic-systems dynamics models, microeconomic decision-theory formulations, discrete-event simulations, and agent-based simulations.

These multilevel models, with associated large, interactive visualizations, are examples of policy flight simulators. The last portion of this chapter shows how these simulators have been used to address the process of transforming healthcare in the United States (Rouse & Serban, 2014). I discuss model-based applications involving scaling and optimization of results of randomized clinical trials; competition and innovation in health ecosystems; and population health involving the integration of health, education, and social services.

Chapter 8: Intelligent Systems

This chapter addresses artificial intelligence (AI), including machine learning. I review the history of both AI and contemporary AI. Several AI models of intelligence are compared and contrasted, particularly in terms of which aspects of intelligence are amenable to which models. This has implications for what can be automated and what should only be augmented.

I pay particular attention to "intelligent interfaces," a phrase I coined to characterize AI systems that literally understand their users (Rouse, 2007). For example, beyond understanding routes and other cars, a driverless car needs to understand its passengers. Such understanding will enable human–AI interaction in terms of information management, intent inferencing, error tolerance, adaptive aiding, and intelligent tutoring. "Explainable AI" is, therefore, possible, which is likely to be key to market success (Rouse & Spohrer, 2018).

I address cognitive assistants, human-built computational processes that rely on machine learning or AI to augment human performance. Cognitive assistants rely on two broad classes of models. One is a model of the domain, for example, aircraft piloting or cancer care. The other is a model of the user in terms of his or her workflows, communications, calendar, and contacts. The examples discussed include aircraft piloting and cancer care.

Chapter 9: Enterprise Transformation

In Chapters 4–8, I discuss product and service markets and competitors, how technology enables and supports competition, how systems failures are addressed, and opportunities in intelligent systems technology. At some point, the changing landscape in these arenas requires changing not just products and services, but the enterprise itself. In this chapter, I address enterprise transformation.

I present a qualitative theory, namely, *enterprise transformation is driven by experienced and/or anticipated value deficiencies that result in significantly redesigned and/or new work processes as determined by management's decision-making abilities, limitations, and inclinations, all in the context of the social networks of management, in particular, and the enterprise, in general.*

With 200 % turnover in the Fortune 500 in the past two decades, needs for fundamental changes are ever increasing and very difficult to address. I present a framework that characterizes the ends, means, and scope of transformation. I summarize numerous examples.

This background provides the basis for a computational theory of enterprise transformation that enables consideration of alternative strategy choices. These choices include predicting better, learning faster, and acting faster. Predicting better decreases uncertainties about future states of market demands and competitors' offering. Learning faster implies that knowledge gained is more quickly incorporated into enterprise competencies. Acting faster quickly turns predictions and knowledge into practice. These three strategy choices each require investments. We would like to know the conditions under which these investments make sense. Also of interest are the conditions under which one should avoid investments.

Chapter 10: Exploring Possible Futures

This chapter integrates the many insights and learnings from earlier chapters. I elaborate and distill the central theme of this book—models are very useful for predicting possible futures rather than a single projected future. Prediction-based insights are the goal. True clairvoyance is extremely rare. One should not expect it.

The many examples and case studies I discuss throughout this book emphasize model-based support for decision-makers. In this chapter, the many lessons from these experiences are integrated. Of particular importance is fostering organizational acceptance of model-based decision support.

I conclude by addressing evidence-based decision-making. For "what is?" decisions, evidence can be gleaned from data. In contrast, "what if?" decisions require predictions ranging from those derived from intuitive gut feelings to those derived from formal computational models. For major decisions with potentially high consequences, the intuitions of experts in comparable situations may be reasonable (Klein, 2003). However, in unfamiliar situations with high consequences, formal consideration of possible futures is undoubtedly prudent.

REFERENCES

Kahneman, D. (2011). *Thinking, fast and slow.* New York: Farrar, Straus and Giroux.

Klein, G. (2003). *Intuition at work: Why developing your gut instincts will make you better at what you do.* New York: Doubleday.

Rouse, W. B. (1976). A library network model. *Journal of the American Society for Information Science*, 27(2), 88–99.

Rouse, W. B. (2015). *Modeling and visualization of complex systems and enterprises: Explorations of physical, human, economic, and social phenomena.* Hoboken, NJ: John Wiley.

Rouse, W. B. (1969). *MOSES: Mission oriented system effectiveness synthesis.* Portsmouth, RI: Raytheon Submarine Signal Division.

Rouse, W. B. (2007). *People and organizations: Explorations of human-centered design.* New York: Wiley.

Rouse, W. B. (2016). *Universities as complex enterprises: How academia works, why it works these ways, and where the university enterprise is headed.* Hoboken, NJ: Wiley.

Rouse, W. B., & Serban, N. (2014). *Understanding and managing the complexity of healthcare.* Cambridge, MA: MIT Press.

Rouse, W. B., & Spohrer, J. C. (2018). Automating versus augmenting intelligence. *Journal of Enterprise Transformation.* doi:10.1080/19488289.2018.1424059

Rouse, W. B., Lombardi, J. V, & Craig, D. D. (2018). Modeling research universities: Predicting probable futures of public vs. private and large vs. small research universities. *Proceedings of the National Academy of Sciences*, 115(50), 12582–12589.

Thaler, R. H., & Sunstein, C. R. (2008). *Nudge: Improving decisions about health, wealth, and happiness.* New Haven, CT: Yale University Press.

CHAPTER 2

Elements of Modeling

This chapter is the most conceptual in the overall book. I discuss the central elements of modeling that underpin the many case studies that I discuss in later chapters. The line of reasoning outlined here provided the foundation for these case studies.

The obvious starting point for this discussion is a definition of the noun "model." The various definitions, resulting from a quick Google search, emphasize (1) an exemplar to be followed or imitated and (2) 3D representations of objects, typically on a smaller scale. Table 2.1 includes these notions as well as several others.

The first four rows of this table elaborate the exemplar notion of a model. The fifth and sixth rows address models of objects. Rows 7 and 8 define models as used in this book.

Representational models depict phenomena. The five examples in Chapter 1 were based on a range of depictions of phenomena—queues, cash flows, control systems, networks of components, and networks of flows. These depictions represent abstractions of physical reality. For some of the case studies in later chapters, I will discuss how multiple abstractions of the same phenomena are composed into an overall representational model.

Computational models enable "solving" representational models in the sense of computing the evolution of the phenomena represented, typically over time. These computed futures are predictions of the evolution of phenomena as a function of assumptions about initial conditions and various parameters within the underlying representational model.

Such outputs are *calculations* of how the representational model will evolve, but not *predictions* of how the real phenomena of interest will evolve. I make

Computing Possible Futures. William B. Rouse. Oxford University Press (2019).
DOI: 10.1093/oso/9780198846420.001.0001

Table 2.1 ***Types of models and their nature and/or use***

Type of Model	Nature/Purpose of Model
Role Model	These are the characteristics and behaviors of the type of person you aspire to be
Model Student	This is what exemplary attitude, behaviors, and performance looks like
Model Home	This is what your home will look like if you buy this home or rent this apartment
Fashion Model	This is what you will look like if you buy this dress, shoes, coat, etc.
Model Car, Ship, or Cathedral	A miniature version of a full-size object of interest, e.g., a vehicle or a building
Product Model	A current or past version of a vehicle, appliance, device, or other product, e.g., 1960 Ford F100 pickup truck
Representational Model	A depiction, e.g., a structural description or a set of equations, that reflects the nature of phenomena of interest
Computational Model	Hardware and/or software code that enables "solution" of a representational model

this distinction to emphasize the difference between the model world and the real world. As illustrated in Chapter 1, the simplifications made to enable the formulation of a representational model can result in predictions that are unlikely to be manifested in reality.

The last row of Table 2.1 indicates hardware and/or software code. Most contemporary modeling tools are software packages that run on digital computer hardware. Many years ago, I employed hybrid computation that involved analog and digital computers, where the electronic circuits on the analog computer would, for example, simulate the differential equations of interest. Hybrid computation is much less common now.

Process of Modeling

This section addresses the nature of modeling, with an initial emphasis on framing the questions that drive modeling efforts—see Figure 2.1. All too often, framing is quickly done and later regretted. I have found that this tendency can be countered by first focusing on identifying the phenomena associated with

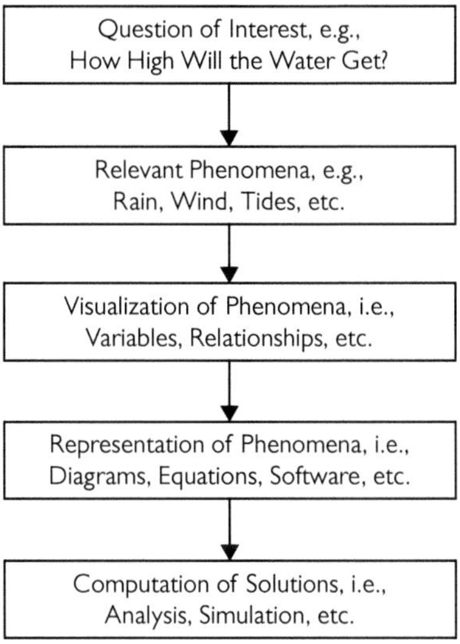

Fig 2.1 Process of modeling.

questions. This involves considering the physical, human, economic, and social phenomena that will need to be addressed to answer the questions of interest.

I have found that the best way to frame the questions of interest is through interaction with the stakeholders in the modeling effort, for example, users, domain experts, and sponsors. Without such interactions, it is quite likely that the engineers and analysts who will develop the model(s) will misunderstand the context of the questions of interest. For example, development of a healthcare-related model would be biased, perhaps even wrong, based on the developers' personal experiences of healthcare.

Relevant Phenomena

With the question(s) of interest defined, at least tentatively, we next need to focus on the phenomena relevant to addressing the question(s). The question might be, "How high will the water get?" The phenomena of interest will likely be rain, wind, tides, and the physical characteristics of the environment, for example, elevations, and structures.

The notion of phenomena has an interesting association with the concept of technology (Rouse, 2015). Arthur (2009) explores the meaning of the term "technology" within the overall concept of technological innovation. He defines technology as a collection of phenomena (physical, behavioral, or organizational) captured and put to use, that is, programmed to our purposes. In many

Table 2.2 ***Class of phenomena versus example phenomena of interest***

Class of Phenomena	Example Phenomena of Interest
Physical, natural	Temporal and spatial relationships and responses
Physical, designed	Input–output relationships, responses, stability
Human, individuals	Task behaviors and performance, mental models
Human, teams, and groups	Team and group behavior and performance
Economic, micro	Consumer value, pricing, production economics
Economic, macro	Gross production, employment, inflation, taxation
Social, organizational	Structures, roles, information, resources
Social, societal	Castes, constituencies, coalitions, negotiations

cases, the phenomena were identified hundreds or thousands of years before they were harnessed for practical purposes, for example, electricity, steam, wind. Elsewhere, I provide a variety of glimpses into these stories (Rouse, 2015).

Table 2.2 summarizes the eight classes of phenomena that I have found useful. A few caveats are in order. First, this taxonomy is not a general taxonomy of all possible phenomena. For example, physicists and chemists would find many of their interests unrepresented. Put simply, the nature of this taxonomy was determined by the phenomena prevalent in the problems and models discussed later in this book.

This table can be used to prompt thinking about the relevance of these classes to questions of interest. The temporal and spatial relationships among rain, wind, and tides are central to predicting how high the water will be as a function of location and time. Human and social phenomena will be central to predicting how a population will respond to water levels, both anticipated and experienced.

Visualization of Phenomena

Visualizing phenomena involves sketching mechanisms underlying phenomena, relationships among phenomena, and plotting relevant data. It is often the case that close inspection of the resulting visualization—a model in itself—leads to the question being answered without deeper modeling. This also enables the engagement of experts on the phenomena, rather than modeling per se.

Fig 2.2 Interaction visualization of the health ecosystem of New York City.

I have many experiences of groups of subject matter experts and key stakeholders exploring large interactive visualizations and gaining important insights from their discussions and debates about what affects what, what trade-offs are crucial, and what leading indicators may portend change. Figure 2.2 shows a large interaction visualization of the health ecosystem of New York City (Yu, Rouse, Serban, & Veral, 2016).

Representation of Phenomena

Representing phenomena involves formulating mathematical and/or computational instantiations of those phenomena meriting deeper study. Composing component models into an overall model involves a variety of issues including consistency of independence, continuity, and conservation assumptions, as well as difficulties of entangled states across models. I discuss representations in more depth below.

Computation of Solutions

Computation involves "solving" the representations to predict the impacts of inputs on outputs. For simple representations, this can sometimes be accomplished with pencil and paper. More complex representations often require computational solutions. Commercial modeling software applications are typically used for this purpose. Often these applications include valuable visualization capabilities. A range of such tools is discussed throughout later chapters.

Alternative Representations

There is a variety of modeling paradigms that can be employed to represent the classes of phenomena in Table 2.2. These alternatives are listed in Table 2.3. Many of the phenomena in the center column of Table 2.3 are rather domain specific.

Model developers can choose from many possible representations. The choice depends on the specific nature of the phenomena, the data available for parameterization of the representation, and the experiences and preferences of the developers. Sponsors will often ask the reasons for choosing particular representations. This is a good question.

Representations usually have a structure, for example, differential equations, and parameters, for example, coefficients. The structure relates to the evolution of the phenomena of interest. The parameters tailor representations to specific contexts. Another common question is the source of data used for parameterization and its validity for this purpose. This is another good question.

Table 2.4 summarizes the key assumptions underlying each modeling paradigm and the typical predictions made using these paradigms. Most of predictions in the right column of Table 2.4 could be applied to a wide variety of problems in many domains, not just those in the center column of Table 2.3.

System State

The notion of system state underlies all the paradigms in Table 2.4. The state of a system is the set of variables, and their current values, sufficient to predict the future states of the system, given knowledge of external forces affecting the system. The above paradigms address state in somewhat different ways.

Dynamic Systems Theory

Differential or difference equations are concerned with continuous time or discrete time, respectively. Continuous time is represented by readings of a traditional analog clock, while discrete time is represented by readings of a digital

Table 2.3 ***Archetypal phenomena and modeling paradigms***

Class	Phenomenon	Modeling Paradigm
Physical, natural	Flow of water	Dynamic systems theory
Physical, natural	Disease incidence/progression	Statistical models, Markov processes
Physical, natural	Cell growth and death	Network theory, biochemistry
Physical, natural	Biological signaling	Network theory, biochemistry
Physical, designed	Flow of parts	Network theory, queuing theory
Physical, designed	Assembly of parts	Network theory, queuing theory
Physical, designed	Flow of demands	Network theory, queuing theory
Physical, designed	Traffic congestion	Network theory, dynamic systems theory
Physical, designed	Vehicle flow	Agent-based models
Physical, designed	Infrastructure response	Dynamic systems theory, network theory
Human, individual	Diagnosis decisions	Pattern recognition, problem-solving
Human, individual	Selection decisions	Decision theory
Human, individual	Control performance	Dynamic systems theory, control theory
Human, individual	Perceptions and expectations	Pattern recognition, Bayes's theory
Human, team/group	Group decision-making	Decision theory, social choice theory
Economic, micro	Investment decision-making	Decision theory, discounted cash flow
Economic, micro	Operational decision-making	Network theory, optimization
Economic, micro	Risk management	Decision theory, Bayes's theory
Economic, micro	Dynamics of competition	Game theory, differential equations
Economic, macro	Dynamics of demand and supply	Dynamic systems theory, optimization
Economic, macro	Prices, costs, and payment	Discounted cash flow, optimization
Social, information sharing	Social networks	Network theory, agent-based models
Social, organizations	Domain social system	Network theory, decision theory
Social, values/norms	Domain values and norms	Network theory, decision theory

Table 2.4 ***Modeling paradigms, common assumptions, and typical predictions***

Modeling Paradigm	Common Assumptions	Typical Predictions
Dynamic systems theory	Newton's laws Conservation of mass Continuity of transport	Response magnitude Response time Stability of response
Control theory	Known transfer function of state transition matrix Stationary, Gaussian stochastic processes Given objective function of errors, control effort	Response time Stability of response Control errors Observability Controllability
Estimation theory	Known dynamics of process Known ergodic (stationary) stochastic process Additive noise inputs	State estimates: filtering, smoothing, prediction Estimation errors
Queuing theory	Known arrival and service processes Future state only depends on current state Given service protocol, e.g., First Come, First Served, priority	Number and time in queue Number and time in system Probability of balk or renege
Network theory	Discrete entities, e.g., agents Decision rules of entities, e.g., agents Typically binary relationships Relationships only via arcs or edges	Shortest distance between any two locations (nodes) Shortest time between any two locations (nodes) Propagation of sentiment among actors
Decision theory	Known utility functions Comparable utility metrics Known payoff matrix Given voting rules	Choice selected Game equilibrium Election results Impacts of incentives
Problem-solving theory	Known human mental model Known information utilization Known repertoire of patterns Known troubleshooting rules	Time until problem solved Steps until problem solved Problem-solving errors
Finance theory	Projected investments Projected operating costs Projected revenues and costs	Net present value Net option value Net capital at risk

clock. Both types of equations are concerned with changes in time. For example, the position of a vehicle in the future depends on its current position and velocity.

Partial differential equations address spatial as well as temporal changes. These representations are particularly useful for flows of liquids (e.g., air and water) across space and time. These types of models are used to predict floods as well as design aircraft wings. The equations are typically solved using discretized numerical approximations.

Control Theory

Feedback is used to modify the control of a system in response to its outputs or, often, in anticipation of its inputs, for example, the road ahead in automobile driving. Control laws are incorporated into the dynamic representation of phenomena to enable predicting the results of variations of the parameters of these control laws. Often, the goal is to optimize the values of these parameters to achieve the best control relative to defined objectives, for example, to minimize path-following errors while also minimizing energy expended to achieve this objective.

Estimation Theory

Thus far, we have assumed that everything is deterministic, that is, the system response is the same every time the same inputs are provided. However, there are often variations, for example, due to bends in the road; wind and rain; other drivers; and so on. These uncertainties can be represented as noise that is added to system state and require the control system to infer or estimate the underlying state based on the observed, noisy state.

Queuing Theory

The state of a system is not always expressed in terms of position, velocity, and so on. For service systems, the state of the system is expressed in terms of number of entities (e.g., people and packages) at each stage of service. In this case, state is discrete while time is continuous. There are probabilistic arrivals of entities needing service and probabilistic durations of service times. Average waiting times and average lengths of queues are usually the primary concerns.

Network Theory

The state of a system may be represented in terms of connections among discrete entities, for example, circuit elements, warehouses, and people. The connections represent flows of, for instance, current, packages, or communications. The entities typically transform inbound flows in some way before sending them on as outbound flows. Such representations can be used to project demands on each entity, determine shortest paths between entities, and predict how communications and sentiments will diffuse through populations.

Decision Theory

State in decision theory can be characterized in terms of the values of attributes that depict alternatives of interest, perhaps across multiple decision-makers or stakeholders. The goal is to maximize expected utility, where use of expected value reflects the probabilistic nature of values of attributes.

Utility functions map attributes values from measured scales to preferential scales. For example, price would be measures in dollars, and performance in the acceleration of a vehicle, for example, the time from 0 to 60 miles per hour. Utility functions would map dollars and seconds to utile values. These mappings are typically nonlinear, as the utility of price drops rapidly for higher prices, and the utility of acceleration would be quite low for 0-to-60 times that impose an unacceptable g loading on the driver.

Decision theory also addresses multiple decision-makers who may be competitors or adversaries. This is the realm of game theory. Typically, there is a payoff matrix that specifies the payoffs to each decision-maker for the collective set of actions of all players. A common objective is to find the conditions under which equilibrium will be reached, such that it is not in anybody's interests to change their decision once they see others' decisions.

Problem-Solving Theory

The state of the system is, for example, the operational availability of the system for use. Problem-solving involves determining why it is not available—diagnosing its failure. Detecting that the system has failed is a decision-theory problem, while compensating for the failure tends to be a control-theory problem.

Problem-solving can be represented with combinations of S-rules and T-rules. S-rules map recognized patterns to diagnostic actions. T-rules employ the structure of the problem to reason through alternative diagnostic actions. Typically, people only employ T-rules when S-rules are unavailable or fail.

Other types of problem-solving include diagnosing underlying reasons for disagreements among stakeholders and may involve inferring underlying utility functions. Much problem-solving research has focused on puzzles such as, for example, the Tower of Hanoi, and games such as, for example, Myst. Puzzles and games allow for much more controlled experimentation.

Finance Theory

State in finance theory is characterized in terms of time series of revenues, costs, and profits. The projected futures of these time series are usually highly uncertain, the more so as the time into the future increases. Discounted cash flow models are used to represent the time value of money, that is, an amount of money received in the future is worth less than that same amount received now due to the interest forgone by having to wait.

Option-pricing models can be employed when there are decision points at which one can exit investments. Owning an option on shares of stock, for example, grants one the right, but not the requirement, to purchase these shares at some point in the future for a specified price. If, when that time arrives, the shares are valued at less than the specified price, one would not exercise the option.

Common Phenomena

The exposition in this section may seem rather complicated. However, there are only a few central concepts—state, dynamic response, feedback control, uncertain states, discrete flows, networks flows, decision trade-offs, systems failures, and economic value. These concepts are laced throughout the examples presented in the remaining chapters of this book. This chapter can serve as a reference when again encountering these concepts.

Summary

This brief excursion into eight modeling paradigms was intended to provide an appreciation for a range of concepts that are employed in the case studies in later chapters. Mastering these eight paradigms would require extensive study and experience, perhaps involving earning one or more academic degrees. Fortunately, mastery is not required to be able to judge whether the approach to modeling being advocated makes sense. Tables 2.3 and 2.4, for instance, could enable one to ask the modelers important questions. In a later section, I discuss what questions are typically asked.

Table 2.5 ***Use of paradigms in later chapters***

Modeling Paradigm	Chapter						
	3	4	5	6	7	8	9
Dynamic systems theory	✔		✔	✔	✔		✔
Control theory	✔			✔		✔	
Estimation theory				✔		✔	✔
Queuing theory				✔	✔	✔	
Network theory		✔			✔	✔	✔
Decision theory		✔	✔	✔	✔		✔
Problem-solving theory				✔		✔	
Finance theory	✔	✔	✔		✔		

Table 2.5 summarizes where these eight paradigms appear in later chapters. I discuss how the paradigms enabled model-based exploration of issues of importance in the domains addressed by each chapter. This catalog of alternative representations is quite rich.

Validating Predictions

Models are used to make predictions to inform answering the questions of interest. These models can be built from first principles, for example, Newton's laws in physics or decision theory from economics. A complementary approach is data-driven modeling from statistics and data science. Combining both approaches can be powerful, particularly when only modest data sets are available to enable purely data-driven modeling.

Validation of predictions is a major issue. Decision-makers will often ask whether a model's predicted outcomes will actually occur. I provide a range of responses to such questions. Validation of prediction-enabled insights raises different issues. Typically, the decision-making group will discuss and debate whether insights make sense or not. This often results in improving the model, while also frequently providing the group with unexpected ideas.

There is an enormous literature on the concept of validity. Within the arena of modeling and simulation, Tolk (2013) provides an interesting and useful exposition. Here are several ways in which model validity is often discussed:

- predictive validity: can the model be employed to accurately predict future system states?
- construct validity: does the structure of the model and data employed reflect what is known about the phenomena being modeled?
- population validity: are the model's predictions generalizable across populations beyond those studied?
- ecological validity: is the model based on appropriate assumptions relative to the phenomena and population being studied?
- face validity: are the model's structure, underlying data, and predictions reasonable?
- insight validity: do people's insights, once discussed and debated, make sense in terms of what is known about the phenomena being modeled?

This book is primarily concerned with face and insight validity. Decision-makers expect models to yield reasonable predictions, as I discuss in the following section. They realize that these predictions represent what *might* happen. Their primary objective is to understand the conditions under which these possible futures might occur. This includes leading indicators of which futures are possibly emerging.

Such conditions and indicators enable decision-making teams to gain insights into the complexity of their broadly defined enterprise. These insights are typically articulated by the decision-making team, rather than by the computational model(s). The team validates these insights by discussing and debating them. This often leads to use of the models to explore new scenarios. Decision-makers want to assure that they deeply understand these insights before they act upon them.

Frequently Asked Questions

I have engaged with well over 100 companies and agencies, working with several thousand executives and senior managers, to develop and employ models to computationally explore possible futures. I discuss some of these experiences in later chapters. At this point, I want to address their reactions to being involved in the process summarized in Figure 2.1.

First and foremost, I have never encountered an executive or senior manager who said, "Well, the model's predictions make no intuitive sense to me at all, but I will go along and make the decisions the model suggests." In such situations,

I had to "peel back the onion" to unearth the assumptions, data, and so on leading to the counterintuitive predictions. This typically resulted in new intuitions being formed or the offending assumptions being changed.

Over hundreds of experiences, I compiled the questions most frequently asked. The first question is "How wrong can I be and still have this decision make sense?" Sponsors of modeling efforts, and users of the models developed, are keenly aware that many assumptions underlie these models. They know that these assumptions cannot possibly be exactly correct. They want to know how wrong the assumptions can be but still result in predictions that are useful to inform decisions.

Their underlying uncertainties include stakeholders' and their intentions. How accurate are the assumptions about customers and competitors? Have we included the right attributes and their relative importance? Do we know the full set of alternatives available to customers and competitors and their attributes? For example, the status quo is often overlooked as an alternative. Consumers might not buy anything.

Savvy sponsors of modeling efforts are naturally skeptical and want to be convinced with compelling and perhaps intuitively reasonable arguments. Once all the models and spreadsheets are done and results digested, decision-makers typically make a "gut check" before they commit. One told me, "You have got to count the numbers right, but numbers are not all that counts."

The second question is "How bad can things get and still have this decision make sense?" Once they have made their decisions and committed resources, underlying uncertainties remain. What are the consequences of these decisions and the implications of these consequences? Decision-makers are fully aware that not all consequences will have been anticipated.

How will stakeholders react to the consequences? Will customers see them as positive, increasing their likelihood of buying? Will competitors identify flaws and develop strategies to exploit them? Will decision-makers have the ability to influence consequences, exploiting positive consequences and remediating negative consequences?

The process of developing and employing models is not an end in itself. Instead, models provide a means for exploration and exploitation of ideas and opportunities and, in general, organizational learning (March, 1991). The people in Figure 2.2 are mutually learning about the health ecosystem of New York City. They also are learning, from each other, from the perspectives of patients, providers, payers, and consultants on the phenomena being explored. They are learning about what *might* happen, as well as the conditions under which these outcomes are likely to happen.

Key Points

This chapter has suggested several important observations:

- Modeling efforts should be driven by the questions that prompted the efforts and the phenomena associated with addressing the questions of interest.
- Visualizations of the phenomena of interest can help both to engage stakeholders and to simplify the trade-offs that need to be addressed.
- There are a variety of representational paradigms that can be employed to model the evolution of the states of the system of interest.
- Choices among representations depend on the specific nature of the phenomena of interest, the data available to support use of this representation, and the expertise and preferences of the personnel involved.
- Computational models—and visualizations—are means to explore phenomena of interest, identify central trade-offs, and foster collaborative solutions.
- Decision-making teams use models to gain insights into the complexity of their broadly defined enterprise with an overarching goal of understanding what might happen and the conditions under which such futures are likely.

REFERENCES

Arthur, W. B. (2009). *The nature of technology: What it is and how it evolves.* New York: Free Press.

March, J. G. (1991). Exploration and exploitation in organizational learning. *Organization Science, 2*(1), 71–87.

Rouse, W. B. (2015). *Modeling and visualization of complex systems and enterprises: Explorations of physical, human, economic, and social phenomena.* New York: Wiley.

Tolk, A. (Ed.). (2013). *Ontology, epistemology, and teleology for modeling and simulation.* Berlin, Heidelberg: Springer-Verlag.

Yu, Z., Rouse, W. B., Serban, N., & Veral, E. (2016). A data-rich agent-based decision support model for hospital consolidation. *Journal of Enterprise Transformation, 6*(3/4), 136–161.

CHAPTER 3

Economic Bubbles

Bubbles eventually burst, whether joyful children playing with soapy water create them, or greedy people playing with other people's money fashion them. In this chapter, I address the higher-education cost bubble, why it is unsustainable, and the ways it will likely burst. I use a computational model of research universities to explore possible futures for these universities.

One of the earliest economic bubbles concerned tulip bulbs in Holland (Dash, 2001). "Tulipomania" involved the speculative buying and selling of rare tulip bulbs in the 1630s by Dutch citizens. Coveted bulbs changed hands for amazingly increasing sums, until single bulbs were valued at more than the cost of a house. When the bubble burst, the value of bulbs quickly plummeted, and fortunes were lost.

We recently experienced a real-estate bubble (Blinder, 2013; Lewis, 2011). In real-estate mortgage markets, impenetrable derivative securities were bought and sold. The valuations and ratings of these securities were premised on any single mortgage default being a random event. In other words, the default of any particular mortgage was assumed to have no impact on the possible default of any other mortgage.

The growing demand for these securities pressured mortgage companies to lower the standards for these loans. Easily available mortgages drove the sales of homes, steadily increasing home prices. Loans with initial periods of low, or even zero, interest, attracted home buyers to adjustable rate mortgages. Many of these people could not possibly make the mortgage payments when the rates were adjusted after the initial period.

Computing Possible Futures. William B. Rouse. Oxford University Press (2019). © William B. Rouse 2019.
DOI: 10.1093/oso/9780198846420.001.0001

This was of less concern than one might think, because people expected to flip these houses by selling them quickly at significantly increased prices. This worked as long as prices continued increasing, but as more and more lower-quality mortgages were sold, the numbers of defaults increased and dampened the increasing prices, which led to further increases of defaults. The bubble quickly burst.

The defaults were not random events, as assumed by those valuing these securities. They constituted what is termed a "common-mode failure," where a common cause results in widespread failure. Thus, these securities were much more risky than sellers had advertised. The consequences of such misinformation were enormous.

Higher Education

Higher education has become the "poster child" for out-of-control costs, replacing healthcare, which now seems more or less under control (Rouse & Serban, 2014). Tuition increases have far outpaced increases of the overall cost of living. This is due to the relative decline of public support for higher education, while administrative costs have been steadily growing much faster than the costs of teaching and research. A primary enabler and consequence of this cost growth has been the creation of student debt levels that exceed the total credit-card debt in the United States.

Out-of-control healthcare costs were not characterized as a bubble because the value of healthcare was not going to suddenly plummet, as the values of tulip bulbs and houses did. Higher education, however, faces the prospects of the bubble bursting if students and their families decide a college education is not worth the cost. Alternatively, and more likely, new business models will result in the "creative destruction" of the traditional academic business model.

I discuss these business models later in this chapter. First, however, let's consider how we got into this situation in the first place. I provide a comprehensive assessment of the situation elsewhere (Rouse, 2016). Here are some highlights.

In the last 40 years, the number of full-time faculty at colleges and universities has grown by 50 %—similar to increases in student enrollment—but, in this same period, the number of administrators has risen by 85 %, and the number of staffers required to help the administrators has jumped by a whopping 240 %. In the 1970s, almost 70 % of faculty were tenured or on a tenure track. Today, that figure is down to 30 %.

Between 1947 and 1995, overall spending increased by 148 %, but administrative spending increased by 235 %. Between 1975 and 2005, the number of administrators and managers increased by 66 % at public institutions, and 135 % at private institutions. A large portion of these administrators includes staff members who support student services, compliance functions, and so on, all of which have grown enormously in recent years. The national average for private colleges is 9 administrators per 100 students: Vanderbilt has 64 administrators per 100 students, Rochester has 40 administrators per 100 students, and Johns Hopkins has 31 administrators per hundred students, corresponding to 1.6, 2.5, and 3.2 students per administrator, respectively.

Combining this cost growth with reduced public support for higher education has resulted in steadily increasing tuition, which is increasing at a rate that is far beyond that of inflation. How can universities get away with these increases, especially during the recent Great Recession? While most organizations tightened their belts during that period, universities substantially increased their girths.

Government-backed student loans have been the key enabler. These types of loans were first offered in the United States in 1958 under the National Defense Education Act and were at first limited to those studying engineering, science, or education. The Higher Education Act of 1965 extended them to all areas of study. Student loan debt has been growing rapidly since the Great Recession, rising to nearly $1.4 trillion by late 2016—as noted earlier, this is greater than the total U.S. credit-card debt.

As a result, college graduates are saddled with debts, on average, of about $30,000 each, and this amount is much higher for professional degrees. Higher education seems to be pretending that government-backed student loans will let them endlessly increase tuition and fees. But the bubble is stretching. Once the terms of people's student loans overrun the terms of their children's student loans, collapse is inevitable.

Market Disruption

Many commentators have suggested that higher education is ripe for market disruption, which perhaps could be caused by online offerings and the unbundling of universities' offerings by nimble competitors. However, the rules of the game for accreditation, and the requirement of accreditation for students to be eligible for government-backed loans, currently make such unbundling infeasible. Such impedances are hindering the pace of transformation, but they will eventually be overcome.

Technology-enabled education is rapidly improving. I recently completed the first lessons of three courses on the best-known MOOC (for Massively Open Online Course) sites:

- Coursera course: "Chicken Behavior & Welfare"
- edX course: "Dinosaur Ecosystems"
- Udacity course: "Design of Everyday Things"

All three courses provide lessons composed of a series of 1–3 minute video clips, interspersed with short exercises and with a multiple-choice quiz at the end. All three have forums where students can interact with the instructor(s) and other students. All three are reasonably engaging, much more so than many traditional lectures.

My sense is that highly polished, well-done MOOCs will increasingly succeed. Once the credentials associated with success in these online courses are acceptable to employers, it is easy to imagine a massive shift away from traditional classrooms. This will be especially likely for professional master's degrees, where distance learning is increasingly common.

Everyone will take the course on any particular topic from the very best instructor of that topic. For example, everyone will take physics from Richard Feynman, and economics from Paul Samuelson. The fact that these luminaries are no longer with us will not be a hindrance. Technology will enable them to teach new developments in their fields, despite never having heard of them during their lives.

Thus, the seeds of market disruption have been sown. How will this disruption play out? Any point predictions will inevitably be wrong. In fact, there are multiple scenarios relevant to the possible futures of higher education. I explore these scenarios and possible futures in the remainder of this chapter.

Four Scenarios

Scenario development should be based on best practices on this topic (Fahey & Randall, 1998; Schoemaker, 1995; Schwartz, 1991). All of the pundits begin by defining the forces that drive the future. There are—at least—four strong driving forces that will affect academia's future:

1. Competition among top universities, both for talent and for resources, will become increasingly intense—there will be a "clash of the titans."
2. Globalization will result in many academic institutions, particularly in Asia, achieving parity in the competition—academia will become "hot, flat, and crowded."
3. Demographic trends portend an aging, but active populace leading to an older student population—higher education will need to become a "lifespan Mecca."
4. The generation of digital natives will come of age, go to college, and enter the workforce—there will be no choice but for academia to become a networked university.

We cannot escape these forces; nor can we fully predict the ways in which they will interact to shape the world over the next 20 years. We can be sure, however, that for academic institutions to compete in this future, their strategies must be sufficiently robust to accommodate these forces. If, instead, they focus on just one scenario—for example, the "clash of titans," which most closely resembles business as usual, although perhaps on steroids—they will almost certainly be at a competitive disadvantage in the future.

Based on the above line of reasoning, I crafted the following four scenarios:

- "Clash of Titans,"
- "Hot, Flat, and Crowded,"
- "Lifespan Mecca," and
- "Network U."

Clash of Titans

I have worked at, consulted with, or served on advisory boards of quite a few top universities. Every one of them pays attention to their *US News & World Report* rankings, as well as other rankings. They aspire to battle with the titans of higher education, and hold their own. This scenario has universities continuing that clash, perhaps clawing their way to higher rankings, albeit in an increasingly competitive environment.

General Description: Academic institutions continue to battle to achieve dominance in various academic disciplines, as well as competing with top universities for overall rankings within the United States, and with premier international universities for global rankings.

Dominant Issues: The competition for talent becomes fierce, with well-endowed chairs becoming the minimum for attracting faculty talent; top students at all levels expect and get near-free education.

Economic Implications: The top players continue to dominate receipt of federal funds, with considerable pushback from other players; costs of facilities and labs soar, and much of the funds for these must be raised from philanthropic sources.

Social Implications: University cultures are sustained, with adaptations for a population that is increasingly diversifying—for both students and faculty—but as ones that are committed to the values and sense of purpose that has been central for recent decades; changing demographics impacts how alumns best relate to their alma maters.

Hot, Flat, and Crowded

Tom Friedman (2005) has argued that the world is flat and we should no longer assume business as usual—his revision of this best seller included a chapter on Georgia Tech and how they transformed education in computing. More recently, Friedman (2008) has argued that the world will be hot, flat, and crowded. In this scenario, academic institutions have to compete with a much wider range of players in a global arena.

General Description: Global parity emerges in graduate education in science and technology, particularly for traditional disciplines and subdisciplines; greater collaboration among institutions emerges; demand for higher education in the United States will nevertheless increase substantially.

Dominant Issues: Many of the best jobs are in Asia; scarcity and constraints dominate sustainability debates; clashes of belief systems create political turmoil and security concerns; meeting demands presents strong challenges.

Economic Implications: Federal and state support diminish as portions of budgets; industrial and philanthropic support are increasingly competitive; sponsors become sensitive to where resources are deployed; undergraduate tuition stabilizes, and increases are less and less acceptable.

Social Implications: Global footprints of top universities increase by necessity; social, cultural, and ethnic diversity of faculty and students increases, in turn; traditional business practices, for example, promotion and tenure, must change to accommodate diversity.

Lifespan Mecca

It is easy—and convenient—to assume that the students of the future will be much like the students of today. However, over the past decade, the number of graduate students 40 years old or older has reached record heights. From 1995 to 2005, the number of postbaccalaureate students aged 40 and older at U.S. colleges and universities jumped 27 %. And during the next two decades, the number of older citizens will rise at even faster rates than the number of those 24 and younger, which suggests that the number of postbaccalaureate students age 40 and over very likely will continue to grow. In this scenario, universities have to address a "student" population with more diverse interests and with expectations that are rather different from those of students of the past and current eras.

General Description: Demand for postgraduate and professional education surges as career changes become quite common; demand steadily grows for education and artistic performances by an increasingly urban older population.

Dominant Issues: Two or three MS or MA degrees become common across careers, as do often required certificate programs; multiple artistic performance and sporting events per day become common at any top university.

Economic Implications: Tuition revenues soar for professional programs and graduate education programs popular with elders; revenues from artistic performance and sports venues become significant portions of university budgets.

Social Implications: The median age of students increases substantially, changing the campus culture markedly; older students in particular expect and get high-quality, user-friendly services; the diversity of faculties increases significantly to satisfy the diversity of demands.

Network U.

Technology is increasingly enabling access to world-class content in terms of publications, lectures, and performances. Higher education can leverage this content to both increase quality and lower costs. This technology has also spawned the generation of "digital natives" who are always connected, weaned on collaboration, and adept at multitasking. In this scenario, academia has to address the different types of students by using very different approaches to delivering education and conducting research.

General Description: Social technology prevails; access to the best content and faculty is universal; nevertheless, students go to college to learn and mature;

however, the classroom experience is now highly interactive, both remotely and face to face.

Dominant Issues: Students and faculty have broad and easy access to knowledge, often via other people; with the "best in class" universally available, local faculty play more facilitative roles in small (10–20) "high-touch" discussion groups.

Economic Implications: More teaching professionals are needed for recitation-sized classes; teaching skills are at a premium; increasing numbers of high-quality programs result in a strong downward pressure on tuition and fees; faculty research becomes near totally externally funded.

Social Implications: Students and faculty are networkers par excellence, both within and across institutions; students' evaluations of teaching effectiveness play an increasing role; students seamlessly transition from K–12, to university and then on to lifespan education.

Computational Model

Figure 3.1 summarizes a computational model of a research university that is elaborated in great detail in my latest book (Rouse, 2016) and was recently

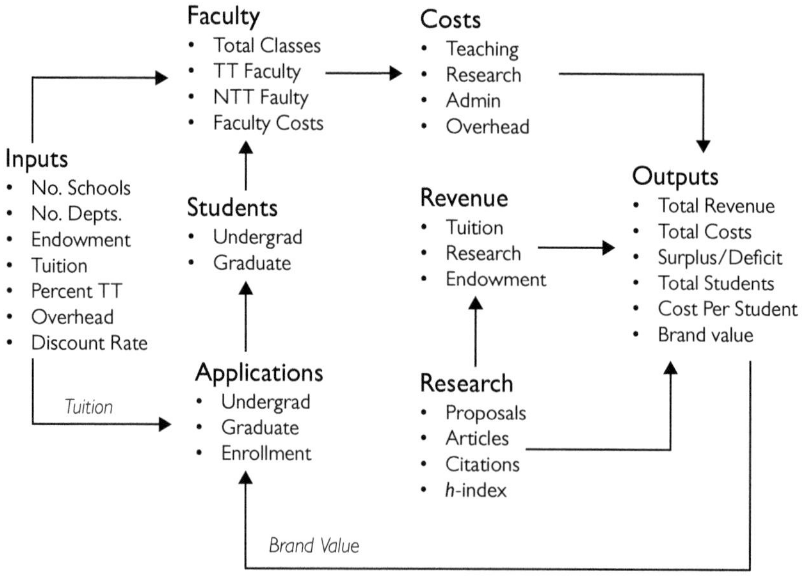

Fig 3.1 Computational model of a research university; NTT: non-tenure-track; TT: tenure-track.

extended (Rouse, Lombardi & Craig, 2018). This model is based on a thorough analysis of a wealth of data pertaining to the various aspects of a university enterprise. This includes sources of funding, alternative publication outlets, predictors of brand value (and hence rankings), workforce structure, administrative practices, and so on.

Student applications are driven by tuition and brand value.[1] Accepted students who enroll, as well as continuing students, determine needs for classes and faculty members to teach these classes, which determines the costs of teaching. Tenure-track faculty members need to pursue research to achieve tenure and promotion. They need to write proposals to attract funding for their research.

The research activities of tenure-track faculty members result in publishing research articles, which are eventually cited and, over time, increase faculty members' h-index, that is, the number of articles cited at least h times. The combination of articles published, citations of these articles, and the h-index, over time, provides an estimate of brand value, which correlates closely with an institution's rankings.

This is all complicated by several phenomena (Rouse, 2016). Research funding is increasingly competitive, with funding decreasing relative to a steadily increasing number of proposals. Publication is increasingly competitive, with opportunities very constrained relative to a steadily increasing number of submissions. The result is that faculty members have to work harder to achieve less success.

The tsunami of competitors is willing to do almost anything to succeed. They will serve long appointments in poorly paid postdoctoral positions with a 15 % chance of securing tenure-track positions. They will work diligently to produce 10, 15, or 20 journal articles before applying for a tenure-track position. In contrast, when I was at that stage of my career, I had *zero* publications.

Revenue comes from tuition, research grants, and endowment earnings. Costs include those for teaching, research, administration, and overhead. Projections of revenues and costs yield model outputs that include various financial metrics plus number of students and brand value. Users of the model are interested in the impacts of various inputs on these model outputs.

Inputs include the number of schools, and the number of departments per school, as these both have a strong impact on administrative costs. Endowments and tuition strongly affect revenues. Overhead rates affect the portion of grants

[1] The various assertions in this section are fully supported by many sources cited and data compilations presented in my previous book (Rouse, 2016).

that can be used for general expenses rather than direct support of research. The discount rate affects projections of the net present value (NPV) of the projected surplus/deficit (S/D).[2]

The percentage of faculty members who are on the tenure track (percent TT) has an enormous impact. Non-tenure-track faculty members teach twice as many classes as tenure-track faculty members do, which substantially reduces costs, especially because non-tenure-track faculty members are usually paid much less than tenure-track faculty members. In addition, tenure-track faculty members spend half their time doing research, which may or may not be paid for from research grants. The publications resulting from this research strongly impact brand value over time.

The trade-off is very clear. Reducing percent TT lowers costs and, in principle at least, decreases tuition. Increasing percent TT increases costs and tuition but enhances brand value. Prospective students seek lower tuition and higher brand value. Leaders of research universities have to decide where to position themselves relative to this trade-off.

There are many other parameters to the model beyond those shown in Figure 3.1. Endowment growth rate and tuition growth rate are also inputs on the "dashboard" for the model. Embedded variables include, for example, administrative salaries and growth rate; initial number of undergraduate and graduate students; growth rates of these populations; and class sizes for both types of students. Such variables are not included on the dashboard. Once they are set for a particular university, they are seldom varied.

There are several submodels within the overall computational model. These submodels relate to finance, administration, research, education, workforce, and brand. The research model projects proposals written, projects funded, articles submitted, and articles published. The workforce model projects decisions about hiring, promotion, tenure, and retirement. The overall model and all the submodels are explained in detail in an earlier book (Rouse, 2016).

Considering the model in Figure 3.1 in terms of the representational paradigms discussed in Chapter 2, three are employed:

- dynamic systems theory: discrete-time difference equations are used to predict the time responses of the input–output relationships;

[2] Net present value is the current value of projected future cash flows, discounted by the interest rate one must forgo or pay due to cash flows being delayed.

- control theory: the size and composition of the workforce is controlled to match the demands for undergraduate and graduate education;
- finance theory: discounted cash flow models are used to calculate NPVs for all financial variables.

Thus, three of the eight paradigms from Chapter 2 are integrated in this computational model of research universities.

Projections for Scenarios

The model was used to project possible futures based on the following assumptions.

Clash of Titans

In this scenario of business as usual on steroids, tuition grows steadily by 5 % annually. Endowment grows steadily by an aggressive 8 %. Percent TT is 80 %, to increase brand value. The percentage of faculty that are tenured after the sixth year is 50 %. The goal is to retain only the most productive faculty members. The undergraduate population grows slowly, at 2 %, while the graduate population grows steadily at 6 %. Administrative costs grow steadily at 6 %, as data shows they have in recent years.

Hot, Flat, and Crowded

With competition among global universities intensifying, graduate enrollment decreases by 4 % annually, reflecting foreign students making different enrollment choices than they did in the past. Having fewer graduate students results in a reduction of percent TT to 30 %. Tuition growth is limited to 2 %, and endowment growth slows to 4 %. The growth of administrative costs is reduced to 3 %.

Lifespan Mecca

Enrollment of older students seeking career changes or pursuing retirement interests results in the graduate population growing at 6 % per year. The undergraduate population grows more slowly, at 2 %. Tuition increases are limited to 2 %, as much of this growth comes from people who are unwilling to pay constantly escalating tuitions. Percent TT decreases to 30 % because the MS and perhaps MA degrees being sought require more teaching faculty. Endowment grows slowly at 4 %. Growth of administrative costs is limited to 3 %.

Network U.

Increased online offerings result in the graduate population growing quickly, at 10 % annually, while the undergraduate population grows more slowly, at 2 %. Classes become small discussion groups; class sizes vary from traditional numbers to much larger. Percent TT decreases to 20 % as the research enterprise becomes more focused on niches of excellence rather than trying to compete across the board. Tuition growth is necessarily limited to 1 % in this highly competitive environment. Endowment grows very slowly, at 2 %, as most alumns have never set foot on campus. Administrative costs necessarily must decline by 5 % annually.

Figure 3.2 portrays the student population at Year 20 and a tuition that achieves an NPV of the S/D that is equal to zero. Note that the tuition does not differ greatly for each scenario. This is due to the model automatically adjusting the number of faculty members to meet demands. This is, of course, easier for non-tenure-track faculty members than for those who are tenured.

The student population is depressed for Hot, Flat, and Crowded as graduate students choose to enroll at globally equivalent but less expensive universities. Lifespan Mecca attracts older American students that swell the graduate ranks. Not surprisingly, Network U. leads to a dramatic growth of online graduate students.

We would expect technology-enabled Network U. to have large classes of remotely connected students, probably very large for lectures and smaller for discussion sections. However, even the discussion classes are likely to be much larger than traditional campus classes. Figure 3.3 shows tuition per semester versus class size, which is expressed in terms of numbers of times greater than the baseline. As in Figure 3.2, this assumes the NPV of the S/D equals zero.

The impact is fairly dramatic. As class sizes increase, the overall model automatically reduces numbers of faculty members, which consequently substantially reduces costs. A rapidly growing student body (see Figure 3.2) while

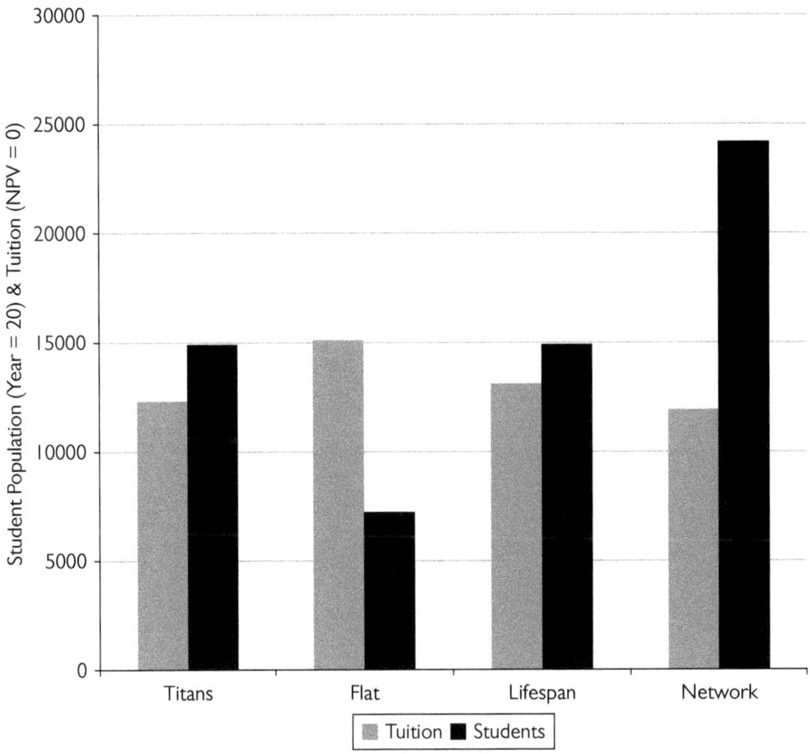

Fig 3.2 Number of students and tuition for each scenario: Clash of the Titans ("Titans"), Hot, Flat, and Crowded ("Flat"), Lifespan Mecca ("Lifespan"), and Network U. ("Network"); NPV: net present value; Titans: Clash of the Titans scenario.

costs of delivery are plummeting enables cutting net tuition from $12,000 per semester to $2,000.

Thus, an undergraduate degree would cost $16,000 in total, assuming it requires eight semesters to earn enough credits to graduate. Of course, by this point, the notion of semesters may be completely obsolete. Pricing will probably be by the course. How courses are bundled will be up to each student. Alternatively, pricing might be by the module, with students mixing and matching the modules to gain the knowledge and skills they seek.

This scenario easily causes one to consider what the university should do with its sizable investment in bricks and mortar. One possibility is that this infrastructure mainly serves the resident undergraduate student population,

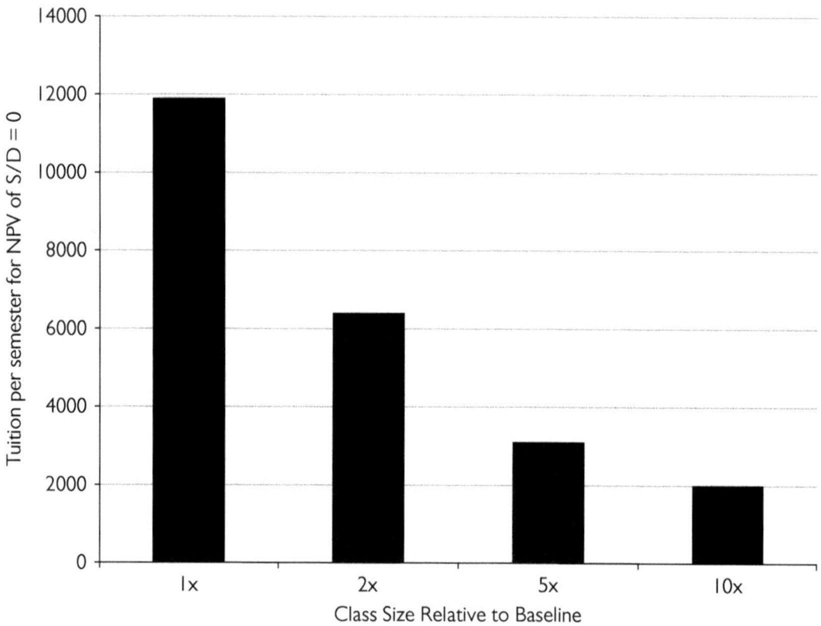

Fig 3.3 Tuition versus class size; NPV: net present value; S/D: surplus/deficit.

while the graduate population needs limited numbers of traditional classrooms and, of course, no dormitories and dining halls.

An overall comparison of these scenarios is as follows:

- Clash of Titans: Begins with slight deficit and then generates a growing surplus as the student population grows; brand value is strong because the percent TT is high;
- Hot, Flat, and Crowded: Leads to declining graduate enrollments and, in later years, steadily increasing deficits; brand value plummets;
- Lifespan Mecca: Leads to strong growth of graduate enrollments and essentially zero deficits; brand value increases, relative to Hot, Flat, and Crowded, because more faculty members are needed to serve increased enrollments;
- Network U.: Leads to exploding graduate enrollments; increasing class sizes, enabled by technology, dramatically lowers costs; brand value steadily decreases as larger class sizes lead to a reduction in faculty size; initial deficits are replaced in later years by huge surpluses.

Implications

You cannot simply choose one of the scenarios. All of them must be addressed, if only to define the early warning signals of their emergence. More strategically, investments in Lifespan Mecca and Network U. constitute hedges against Hot, Flat, and Crowded.[3] A balanced investment portfolio across all scenarios is likely to be the best approach. It will mean that one cannot put all the eggs in the Clash of Titans basket, as that could be quite risky.

Clash of Titans presents a particularly difficult challenge. The current success model at most research universities requires faculty members to work harder and harder to achieve less and less success. Universities need to broaden their views of "gold-standard" research sponsors beyond the National Institutes of Health and the National Science Foundation to include other first-rate sponsors such as the National Aeronautics and Space Administration and the Office of Naval Research. Private foundations and industry sponsorship should be increased.

Universities also need to broaden their views of "gold-standard" journals beyond current "A" journals. They should emphasize citations rather than impact factors, the irrelevance of which is elaborated elsewhere (Rouse, 2016). A paper that earns 100+ citations in a low-impact factor journal should be seen as a home run, not something to be dismissed.

University presidents, provosts, deans, and promotion and tenure committees need to communicate these changes to their faculties, particular junior faculty members. If everyone continues to pursue the old success model, there will be a lot less success, leading to pervasive frustration of junior faculty, and much waste of human and financial resources.

Many universities have envisaged keeping Hot, Flat, and Crowded at bay by creating global campuses, the idea being that those who eschew matriculating in the United States can earn the same credential in Dubai or Singapore. There are merits to this idea, but also limits.

I have experienced many faculty members of foreign extraction advocating the launch of a new campus in their native country. Campus leadership has encouraged this to such an extent that the talent on the home campus was diluted. Having a branch of Carnegie Mellon University (CMU) or MIT in

[3] Hedges, in terms of option-pricing models, are discussed in depth in Chapter 5. These models can be used to project the economic value of alternative investments to hedge the four scenarios.

every country is inevitably unsustainable, particularly in terms of brand value and quality of education.

On the other hand, making a Network U. version of CMU and MIT globally accessible makes much more sense. An interesting hybrid involves pursing a year or two online and the rest of the degree on campus. The key is for the university to make the investment to assure high-quality online offerings that lead to the advertised knowledge and skills. This is not simply a matter of putting one's PowerPoint slides on the web. Proactive engagement of students in the learning experience requires that educators design this experience, monitor its evolution, and constantly improve it.

Lifespan Mecca requires careful attention to what students—ranging from mid-career professionals to eager-to-learn retirees—want and need to gain to achieve their educational aspirations, for example, promotions, new jobs, or simple mastery of history, music, or political science. Many traditional faculty members do not like to teach professionals and see history, music, and political science as "service courses." Success in this arena, therefore, may mean many fewer traditional faculty members.

Comparative Study

The above conclusions are rather general. How do they hold up for particular universities? In pursuit of this question, we explored a subset of the above scenarios for four specific universities using 2016 data from the Center for Measuring University Performance (CMUP; Lombardi, Phillips, Abbey, & Craig, 2016) as well as publicly available data from each institution. The four universities were drawn from the top 160 research universities that CMUP tracks—two from the top of this list, and two from the bottom. In each pair, one was a private institution and one was a public institution. The detailed results are reported in "Modeling research universities: Predicting probable futures of public vs. private and large vs. small research universities" (Rouse, Lombardi & Craig, 2018).

Large versus small is not a distinction regarding number of students. Instead, this distinction concerns annual federal research monies expended plus endowment earnings. Large institutions, as defined here, have substantial financial resources but not necessarily large student bodies. Small institutions are at the low end of the 160 top research institutions. To put this in perspective, there are almost 2,300 nonprofit institutions of higher education in the United States.

Three scenarios, derived from the scenarios discussed earlier, were of particular interest. They may work independently but also may have combined effects on projected results:

- S1: Competition for federal dollars and publication in top journals is steadily increasing. The current success model at most research universities requires faculty members to work harder and harder to achieve less and less success, with proposal writing consuming increasing amounts of time, and publication preparation receiving decreasing attention.
- S2: Foreign-student applications to graduate programs have decreased in recent years, due to competition from other countries and, more recently, concerns about U.S. immigration policies. These professional master's degrees are typically "cash cows" for research universities, subsidizing many other aspects of the enterprise.
- S3: Highly polished, well-done MOOCs will increasingly succeed. Once the credentials associated with success in these online courses are acceptable to employers, it is easy to imagine a massive shift away from traditional classrooms for some categories of students, especially those seeking professional credentials and master's degrees where distance learning is already recognized and increasingly common.

The three scenarios are succinctly defined as follows: (1) S1: status quo; (2) S2: graduate student population declines by 5 % annually; (3) S3: $10K: graduate tuition declines to $10,000, due to online offerings. Class size in S3 was varied from baseline to 10X or 1,000, creating three instances of S3 (S3: $10K, S3: 10X, and S3: 1K) rather than adding a fourth and fifth scenario. This reflects that fact that the external competitive driver is the same in all three cases. What differs is the institution's response to the scenario.

The model was carefully fit to each university's CMUP and public data. Computational results were then explored for each institution individually. These results show that S3: $10K is the worst scenario, resulting in a negative NPV of the S/D for all four universities, because the number of students does not decrease while revenue decreases substantially.

Three of the cases—S2, S3: 10X, and S3: 1K—lead to substantially reduced numbers of faculty, which undermines institutional publishing productivity and, hence, brand value. S3: 1K is the most profitable because the number of students does not decrease but faculty numbers are cut by over 90 %. Brand value, of course, plummets but only in a relative manner.

Institutions with significant resources are simply not going to let these futures happen to them. High-resource institutions have been the "first movers" in enabling S3: $10K. Thus, they are cannibalizing their professional master's degree "cash cows" before others do. They are likely to become the infrastructure platforms and perhaps content providers for resource-poor institutions. This raises the possibility that these resource-poor institutions will disappear or be absorbed by others.

How do the different scenarios affect the four institutions studied? Brand value decreases, due to diminishing returns from research sponsors, and this affects all institutions similarly. The ratios of the brand values of large institutions to those of small institutions range from 4.7 to 7.7 across the scenarios. Thus, the top-ranked institutions will likely remain on top. The substantially declining research productivity of all four institutions should be a major governmental concern in terms of economic development, national security, and so on.

The change in NPV differs significantly across large and small institutions, particularly for S3: $10K. The two large institutions average NPV = –$2.8 billion, while the two small institutions average NPV = –$2.3 billion across a 20-year time horizon. The Year 20 revenues for the large institutions average $6.8 billion, while those for the two small institutions average $1.0 billion. Clearly, the small institutions are not in a position to weather such losses, due to the S3: $10K technology infusions transforming their "graduation education" business. The lower-ranked members of these highly ranked 160 research universities will face significant challenges.

In late 2018, we projected that these dire consequences would begin to emerge over the next few years. Much to our surprise, they began to emerge over the next few months, rather than years. S1 marches on incrementally, but evidence for S2 and S3 has been startling. Foreign-student applications have steadily decreased, due to the increasing quality of institutions in other countries and to immigration worries. High-quality, low-priced online programs for professional graduate degrees are attracting many corporate sponsors away from traditional programs. For the past few years, an average of 10 institutions have been closing each year, with this average expected to soon increase to 15 per year. The bubble may not have yet burst, but the balloon certainly has a serious leak.

Executives' Reactions

I have demonstrated this model to a range of presidents, provosts, and deans at research universities. The computational model and documentation has been provided to roughly 15 universities. The model is driven by data sets that are readily available to anyone, so university executives accept the validity of Figure 3.1 and the model parameters gleaned from these data sets.

However, many of them are quite unhappy with the conclusions reached using the model. These conclusions challenge the traditional model of academic success. This challenge suggests that smaller research universities should no longer chase the big players such as Michigan and Stanford. They will inevitably become unable to afford to compete in this way.

The unhappiness of these executives seems to stem from their clinging to the model of academic success that enabled their success, which typically occurred decades earlier. As teachers, they are inclined to mentor research students to follow in their footsteps. However, these footsteps lead to students aspiring to positions that are steadily disappearing. For all but the most highly resourced institutions, the academic game is inevitably changing.

Key Points

- We cannot predict what mix of these scenarios will actually emerge, although the forces driving these changes are already evident.
- Universities need strategies and investments that enable robust responses to whatever mix of scenarios emerges.
- The higher-education cost bubble will inevitably burst, probably facilitated by increasingly powerful and sophisticated technology platforms.
- Universities need the right portfolio of investments in the hedges that will assure success despite the bursting of the bubble.
- Without such changes, many institutions of higher education will disappear amidst this "creative destruction" (Schumpeter, 1942).

REFERENCES

Blinder, A. S. (2013). *After the music stopped: The financial crisis, the response, and the work ahead*. New York: Penguin.

Dash, M. (2001). *Tulipomania: The story of the world's most coveted flower and the extraordinary passions it aroused*. New York: Broadway Books.

Fahey, L., & Randall, R. M. (Eds.). (1998). *Learning from the future: Competitive foresight scenarios*. New York: Wiley.

Friedman, T. L. (2005). *The world is flat*. New York: Farrar, Straus and Giroux.

Friedman, T. L. (2008). *Hot, flat, and crowded*. New York: Farrar, Straus and Giroux.Lewis, M. (2011). *The big short: Inside the doomsday machine*. New York: Norton.

Lombardi J., Phillips, E. D. C., Abbey, C. W., & Craig, D. D. (2000–2016). *The top American research universities: Annual report*. Tempe, AZ, and Amherst, MA: Arizona State University and University of Massachusetts Amherst.

Rouse, W. B. (2016). *Universities as complex enterprises: How academia works, why it works these ways, and where the university enterprise is headed*. Hoboken, NJ: Wiley.

Rouse, W. B., & Serban, N. (2014). *Understanding and managing the complexity of healthcare*. Cambridge, MA: MIT Press.

Rouse, W. B., Lombardi, J. V, & Craig, D. D. (2018). Modeling research universities: Predicting probable futures of public vs. private and large vs. small research universities. *Proceedings of the National Academy of Sciences*, *115*(50), 12582–12589.

Schoemaker, P. J. H. (1995). Scenario planning: A tool for strategic thinking. *Sloan Management Review*, *36*(2), 25–40.

Schumpeter, J. A. (1942). *Capitalism, socialism and democracy*. New York: Harper & Brothers.

Schwartz, P. (1991). *The art of the long view: Planning for the future in an uncertain world*. New York: Currency Doubleday.

CHAPTER 4

Markets and Competitors

In the last chapter, I considered human participants in an economy. In this chapter, I address humans as consumers of products and services as well as participants in designing these products and services. We are interested in the humans that invest in new offerings, judge their effectiveness, design and develop these offerings, operate and maintain these offerings, and consume these products and services. In general, I term these humans "the stakeholders" in these offerings.

We had developed aiding and training systems for a variety of sponsors, ranging from the U.S. Air Force, Army, and Navy, to the National Aeronautics and Space Administration, to a range of electric utilities and the maritime industry (I discuss a few examples in later chapters). Our clients often asked about our design methodology that yielded these systems. Could we teach their technical staffs to think as we did?

These clients asked us to help them determine what their markets wanted and how they could meet these needs better than their competitors. They wanted a systematic model-based approach to making these determinations. Boeing, Digital Equipment, Honeywell, Lockheed, Motorola, Raytheon, and 3M were soon customers. Not long after, Hitachi, Rolls Royce, Rover, and other international organizations became customers.

Our customer base became roughly 20 large technology-oriented companies and agencies. We grew the company more by expanding across divisions of these enterprises than by recruiting totally new customers. As I recall, about 80 % of our sales each year came from this customer base.

Our model-based approach was based on a construct I developed and termed "human-centered design." This approach addresses the concerns, values, and

Computing Possible Futures. William B. Rouse. Oxford University Press (2019). © William B. Rouse 2019.
DOI: 10.1093/oso/9780198846420.001.0001

perceptions of all the stakeholders in designing, developing, manufacturing, buying, and using products and systems. The basic idea is to delight the primary stakeholders and gain the support of the secondary stakeholders.

This notion first occurred to me at a workshop in the late 1980s at the NASA Langley Research Center near Hampton, Virginia. Many participants were discussing pilot-centered design that focused on enhancing aircraft pilots' abilities, overcoming pilots' limitations, and fostering pilots' acceptance. I suggested that we should do this for all the human stakeholders involved in the success of an aircraft program. People asked what I specifically meant.

I responded, "Pilots may fly 'em, but they don't build 'em or buy 'em!"

In other words, pilots being supportive of design choices may be necessary for success, but it is not sufficient. The airlines have to want to buy the airplanes, and the aerospace companies have to be willing to produce them. The buyers and the builders have criteria beyond those important to pilots.

I elaborated the human-centered design construct in a series of workshops with clients that resulted in a book, *Design for Success* (Rouse, 1991). Two other books soon followed (Rouse, 1992, 1993). The human-centered design methodology has been applied many times and continually refined (Rouse, 2007, 2015).

The premise of human-centered design is that the major stakeholders need to perceive products and services to be valid, acceptable, and viable. Valid products and services demonstrably help solve the problems for which they are intended. Acceptable products and services solve problems in ways that stakeholders prefer. Viable products and services provide benefits that are worth the costs of use. Costs here include the efforts needed to learn and use products and services, not just the purchase price.

Product Planning

Once the books were provided during the workshops, participants had another suggestion. One participant put it crisply, "We don't really want to read these books. We would like tools such that in using the tools we would be inherently following the principles in the books." Another participant said, "We don't just want knowledge; we want executable knowledge."

We agreed with the idea, but we were very slow getting started. Finally, two customers, independently, offered to buy corporate-wide licenses for the tools

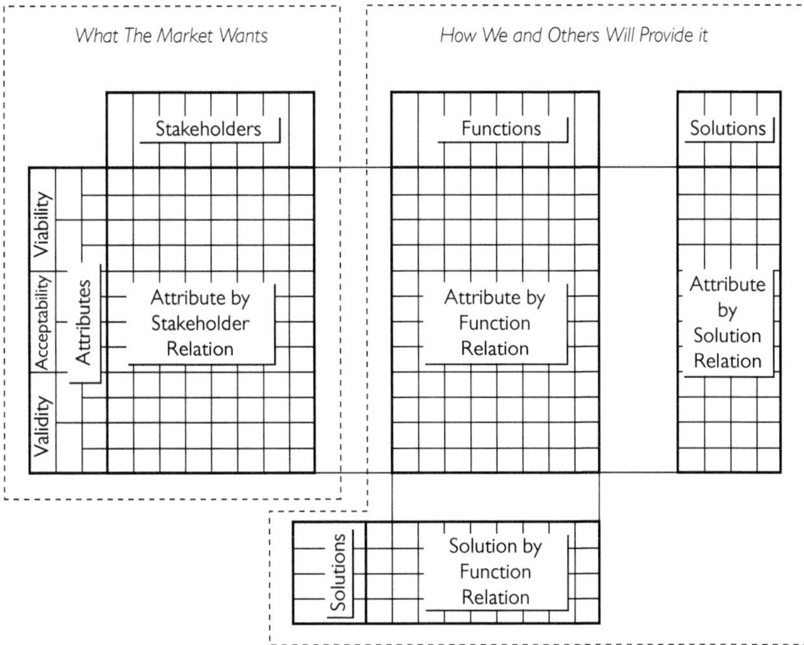

Fig 4.1 Model structure of the *Product Planning Advisor.*

even though they—and we—did not know what they were buying. The initial payments toward these licenses provided the resources to get started.

The first tool was the *Product Planning Advisor*. This tool embodied the principles of human-centered design, built around multistakeholder, multiattribute utility theory (Keeney & Raiffa, 1976) and quality function deployment (Hauser & Clausing, 1988)—see Figure 4.1. This eventually became our best-selling tool (Rouse & Howard, 1993), but it did not get there smoothly.

We formed user groups at the two companies that had committed resources and asked everyone in both groups what they wanted the tool to do. We used this list of desires to build a tool that provided functions and features that *anybody* had requested. When we demonstrated the prototype, virtually every user was overwhelmed. We had provided what they wanted, but it had too many options, modes, and so on.

We went back to the drawing board and redesigned the *Product Planning Advisor* to only provide the functions and features that *everybody* had requested. This version was a success (see Figure 4.2) and we eventually sold many hundreds of copies to over 20 well-known companies. The *Product Planning Advisor* was sold in conjunction with training workshops where participants learned

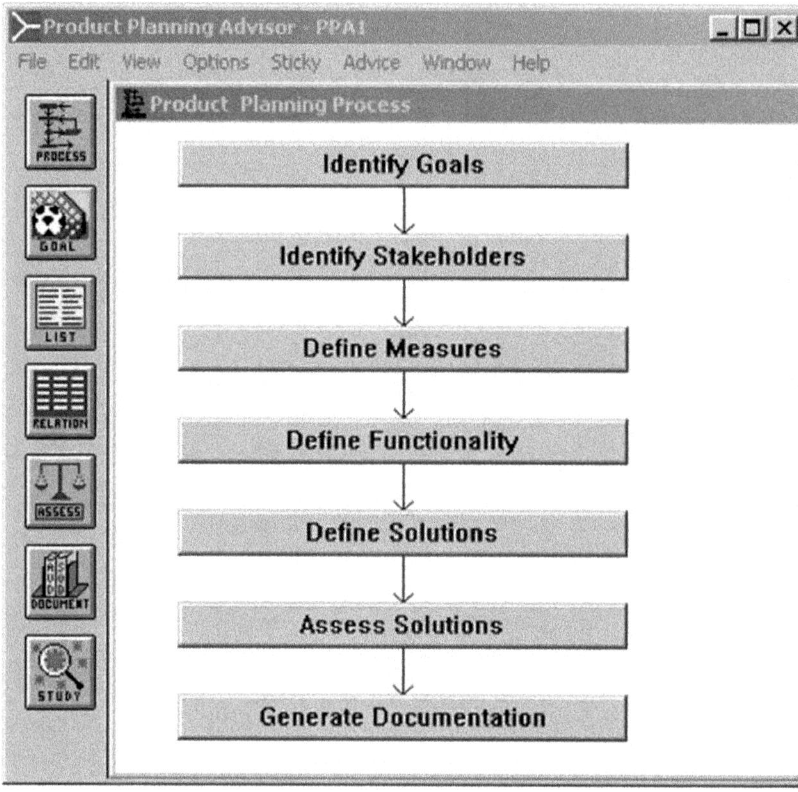

Fig 4.2 The *Product Planning Advisor.*

human-centered design and how to use the *Product Planning Advisor*, all in the context of real product planning problems of importance to their companies.

The *Product Planning Advisor* is used to create and manipulate market and product models to explore the overall utility of both your current and your anticipated offerings to the marketplace, as well as your competitors' offerings. The "What the Market Wants" section of the model characterizes the stakeholders in the product or service and their utility functions associated with context-specific attributes clustered in terms of validity, acceptability, and viability.

The *Product Planning Advisor* is used to create the elements of each stakeholder's multiattribute utility function given by

$$U_i = w_{i1}u_{i1} + w_{i2}u_{i2} + \cdots + w_{iN}u_{iN} \tag{4.1}$$

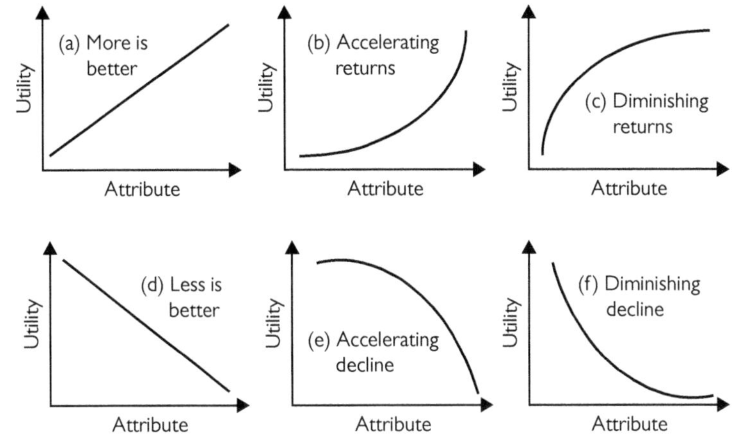

Fig 4.3 Typical forms of utility functions.

where the w's are weights and the u's are utility functions that can be expressed in a variety of forms, for example, linear increasing, linear decreasing, and diminishing returns, as shown in Figure 4.3. Thus, the w's and u's are particular to one stakeholder or class of stakeholders.

Figure 4.1 assumes a common set of attributes across all stakeholders but, obviously, not everyone cares about everything, so for each stakeholder, many of the w's are zero. For example, the stakeholders might be the buyer of a product, the users of a product, and the maintainers of a product. The buyer will, of course, be concerned with the price of the product. The other two stakeholders may not be concerned with price.

The utility functions defined by Equation (4.1) are combined into an overall utility using the following equation:

$$U = W_1 U_1 + W_2 U_2 + \cdots + W_M U_M \tag{4.2}$$

While the w's denote how much a given stakeholder cares about a particular attribute, the W's denote how much we care about each stakeholder's preferences.

The section of Figure 4.1 labeled "How We and Others Will Provide It" specifies, on the right, the attribute values associated with each solution. The functions associated with each solution are defined on the left of this section. Functions are things like steering, accelerating, and braking, as well as functions that may not be available in all solutions, for example, a backup camera.

Attribute-to-function relationships are expressed on a scale from −3 to +3. Positive numbers indicate that improving a function increases the attribute. Negative numbers indicate that improving a function decreases an attribute. For example, a backup camera may increase the price of the vehicle but decrease insurance costs. The use of these relationships is elaborated below.

With *N* attributes and *M* stakeholders, there are often hundreds of relationships possible in Figure 4.1. For example, 40 attributes and 5 stakeholders are fairly typical. This scale necessitated the linear models in Equations (4.1) and (4.2). Keeney and Raiffa (1976) discuss more elaborate forms, for example, multilinear forms. However, with 40 attributes, consideration of preferential interactions among attributes would be overwhelming.

Considering the model in Figure 4.1 in terms of the representational paradigms discussed in Chapter 2, three are employed:

- decision theory: multistakeholder, multiattribute utility theory is used to represent the preferences *of* each stakeholder, or class of stakeholders, as well as the decision-maker(s)' preferences *for* each stakeholder;
- network theory: a graph theory model is used to represent relationships between stakeholders and attributes, as well as attributes and functions, thereby enabling the determination of where functional changes can enhance stakeholder satisfaction;
- finance theory: discounted cash flow models are used to calculate NPVs for all financial variables.

Thus, three of the eight paradigms from Chapter 2 are integrated in this computational model of markets and competitors.

Using the *Product Planning Advisor*

Once the model defined by Figure 4.1 is created, the planning team can use the *Product Planning Advisor* to manipulate the model in several ways. The *w*'s and *W*'s can be varied—on a dashboard that displays them all in one place—to assess the sensitivity of *U* to such changes. Typically, people are concerned with how the rank ordering of solutions, when ranked by *U*, changes with such variations. In particular, what combinations of *w*'s and *W*'s lead to competitors' offerings having higher *U*?

One solution is often a strong competitor, namely, the status quo. In other words, customers may not buy anything. The status quo is compelling because the customer already has it, knows how to use it, and it requires little if any additional expenditures. I have found it quite interesting to see planning teams' reactions when they learn that their solution is worse than having nothing.

When teams encounter this situation or, more often, find the competitors' solutions to be superior to theirs, they will use the *Product Planning Advisor's* "How to Improve" function to explore variations of the attributes and functions of their solutions. They can manipulate individual attributes or clusters of attributes to project the impacts of improvements of 10 %, 20 %, and so on.

For any variations they find promising, the *Product Planning Advisor* will use the attribute-to-function relationships, ranging from −3 to +3, to suggest where the improvements being entertained could best be accomplished. This usually leads the team to consider the feasibility and costs of functional changes. In this way, the *Product Planning Advisor* guides the team's thinking, but insights and ideas come from the team.

The hundreds of engagements with clients using the *Product Planning Advisor* always involved multidisciplinary teams, typically from marketing, engineering, manufacturing, finance, sales, and customer support. Clients often reported that planning sessions using the *Product Planning Advisor* improved their team's mental models in the ways discussed in Chapter 6. They now knew what mattered to each functional area and how these concerns traded off across the preference space embodied in the *Product Planning Advisor*.

Much to our surprise, the services associated with the *Product Planning Advisor* and our other tools went well beyond training. We were repeatedly asked to facilitate workshops associated with new product planning endeavors, despite most of the workshop's participants having been trained earlier. I asked a senior executive why these services were continually needed. His response was, "I am not at all concerned with the costs of your software and services. I am totally concerned with the overall costs of success. Your involvement lowers those costs." Having facilitated hundreds of product planning workshops across many industries, we could share countless lessons learned. These cross-industry perspectives were highly valued.

Applications

We conducted a very large number of new product planning engagements with a wide range of enterprises. In this section, I highlight four of these experiences. Before discussing these cases, it is interesting to note a few observations from other cases.

We worked with Rover on the initial conceptual design of the Mini Cooper, before Rover was bought by BMW, who then brought the Mini Cooper to market. We considered four stakeholders: young women, young men, young couples, and young couple with children. The design differences for each stakeholder were interesting. For example, the back seat plays a different role for couples with children. Young women and men differ in dashboard preferences.

An engagement with the government of Singapore focused on the viability of a very large unmanned aircraft to perform reconnaissance and surveillance over this island country. The public was a major stakeholder, but it was difficult to specify what the public wanted. A key insight was that the public wanted this airborne platform to succeed in the missions for which it was commissioned. Hence, these three missions were included as stakeholders in the *Product Planning Advisor's* embodiment of Figure 4.1.

Working with an aircraft engine company, we discovered that engineering and marketing disagreed about the units of measure of customers' key variables. This emerged because the *Product Planning Advisor* asks users to provide units of measure for all attributes. This was surprising, because the leaders of these two functions had worked together for many years. They commented that they had never had any reason in the past to discuss units of measure.

A semiconductor company was listed in the *Guinness Book of Records* for the speed of their microprocessors. Every product planning engagement with them included the objective that they retain their place in the record book. This was held to be of highest importance, even for applications where the increased speed of the microprocessor provided no benefits to customers, due to limitations of the other elements of the system. They later relented on this objective, as one of the cases below discusses.

We worked with an Israeli chemical company planning new pesticides and herbicides. They were required to test these products on rats to assure that they were not carcinogenic. They reported that none of the rats had developed cancer because all of them had died immediately upon ingesting the chemicals. However, this was inconsistent with the required testing protocol. This is one of the most unusual product planning engagements I ever experienced.

A British aerospace company acquired our product planning methods and tools. They were concerned about product support. We guaranteed them that any problem encountered or question that emerged would be solved or answered within 24 hours. We met this commitment, and the managing director, at a public-industry meeting, commented that this quality of service was amazing and that he had never before experienced such responsiveness. The Internet and email enabled all of this.

The following four examples of how the Product Planning Advisor has been used illustrate the ways in which this tool is applied and the types of insights that are gained. In particular, these examples depict trade-offs across stakeholders and how the impacts of assumptions can be explored. It is important to note that these examples show how product planning teams have reached counter-intuitive conclusions using the *Product Planning Advisor*. However, use of the *Product Planning Advisor* does not, by any means, always result in such conclusions.

Automobile Engine

A team working on new emission control systems decided to evaluate an earlier technology investment using the *Product Planning Advisor*. They compared the chosen approach to four other candidates that had been rejected with the earlier decision. Development and use of the market/product models resulted in the conclusion that the chosen approach was the worst among the five original candidates.

This surprising conclusion led to an in-depth exploration of the assumptions built into their *Product Planning Advisor* models. This exploration resulted in further support for these assumptions. Reviewing these results, the team leader realized that the earlier decision had not fully considered the impact of the alternatives on the manufacturing stakeholder. The earlier choice had been of high utility to customers and other stakeholders but would have been very complex to manufacture. As a result of this insight, a new approach was adopted.

Microprocessors

A major competitor in the semiconductor market was planning a new high-end microprocessor. They were very concerned with time to market, worried that their next generation might be late relative to the competition. Their planning team included people from engineering, manufacturing, marketing, and finance. Using the *Product Planning Advisor*, they found that time to market was critical, but it was not clear how it could be significantly decreased.

One of the manufacturing participants suggested a design change that, upon analysis, would get them to market a year earlier. The team adopted this suggestion. He was asked, "Why have you never suggested this before?" He responded, "Because you have never invited manufacturing to these types of meetings before." Product planning with the *Product Planning Advisor* often results in the involvement of a richer set of internal stakeholders.

Digital Signal Processor

The product planning team began this effort, convinced that they already knew the best function/feature set with which to delight the market. The marketing manager, however, insisted that they test their intuitions by using the *Product Planning Advisor*. After developing the market/product models and using them for competitive analyses, the team concluded that assumptions regarding stakeholders' preferences for three particular attributes, as well as the values of these attributes, were critical to their original intuitions being correct.

Attempts to support these assumptions by talking with stakeholders, especially end users and customers, resulted in the conclusions that all three assumptions were unsupportable. The team subsequently came back to the drawing board and pursued a different product concept.

Medical Imaging System

A product planning team had developed an advanced concept for medical imaging that they argued would enable their company to enter a very crowded market, where a couple of brand-name companies currently dominated. They used the *Product Planning Advisor* to assess the market advantages of their concept relative to the offerings of the market leaders. Initial results showed a considerably greater market utility for their potential offering. Attention then shifted to the likely reactions of the market leaders to the introduction of this advanced product.

The team's expectation was that the leaders would have to invest in two years of R & D to catch up with the new technology embodied in their offering. However, using the "How to Improve?" feature for *Product Planning Advisor* models of the competitors' offerings resulted in the conclusion that the best strategy for the market leaders was to reduce prices significantly. The team had not anticipated this possibility—someone said, "That's not fair!" This caused the team to reconsider the firmness of their revenue projections, in terms of both number of units sold and price per unit.

Summary

These four examples serve to illustrate several types of issues in new product development. The first example showed how the concerns of a secondary stakeholder could affect the attractiveness of a solution. The second example illustrated how a planning team gained insights via the discussions and debates that this tool engenders. The third example depicted the impact of unsupportable assumptions regarding the preferences of primary stakeholders. The final example demonstrated how the likely reactions of competitors impact the possible market advantages of a product or system. Taken together, these four examples clearly illustrate how a human-centered orientation helps to avoid creating solutions that some stakeholders may want but other stakeholders will not support or buy.

Advisor Series

Two other expert system-based tools followed the *Product Planning Advisor*—the Business Planning Advisor and the Situation Assessment Advisor. These two tools provide examples of technical success but market failure compared to the Product Planning Advisor.

Several of our clients asked us to assess their business processes. They liked our product planning process and wondered if we could help them by creating a similar tool for business planning. The *Business Planning Advisor* was the result. It included a rule-based expert system that would assess a business plan in term of eight attributes and then project the likelihood of success.

This tool faced a very competitive market where there were many business-planning tools. Beyond the expert system embedded in the *Business Planning Advisor*, our tool was not that different from the others. The *Business Planning Advisor* was technically sound but did not yield the reactions—or sales—experienced with the *Product Planning Advisor*.

Nevertheless, we sold many business-planning engagements. These experiences led us to realize that many clients did not really understand their market situations. Fairly often, we encountered perceptions that may have reflected past glories but did not capture current opportunities and threats. This led us to develop the *Situation Assessment Advisor*, a rule-based expert system that assessed market situations based on users' answers to a large set of rather specific questions. The *Situation Assessment Advisor* would assess the likelihood that an enterprise was in or headed to one or more of 10 classic market situations.

An executive team at their annual planning offsite typically used the *Situation Assessment Advisor*. It was well received. However, a company would buy one copy to use once per year. In contrast, they would buy 20–40 copies of the *Product Planning Advisor* and use them frequently. The *Situation Assessment Advisor* was a technical success but a market failure. We learned to focus on supporting tasks that are performed frequently by many people.

I discuss a fourth tool, the *Technology Investment Advisor*, in Chapter 5. We marketed the *Product Planning Advisor*, the *Business Planning Advisor*, the *Situation Assessment Advisor*, and the *Technology Investment Advisor* as the *Advisor Series* of planning tools, which is summarized in Figure 4.4. The basic idea was that companies would use the *Situation Assessment Advisor* to assess their market situation, which would inform using the *Business Planning Advisor* to formulate their business strategy, which would drive their use of the *Product Planning Advisor* to plan product and service offerings, which would be informed by the technology investment portfolio formulated with the *Technology Investment Advisor* to enable these offerings.

This provided a pretty compelling marketing story, but the *Product Planning Advisor* and the *Technology Investment Advisor* dominated our engagements with clients. This was due, at least in part, to these tools being more sophisticated in terms of explicit models rather than more opaque expert systems. In other words, with the *Product Planning Advisor* and the *Technology Investment Advisor*, users created models that they could then use to explore various scenarios. With the *Situation Assessment Advisor* and the *Business Planning Advisor*, the underlying models were "hardwired."

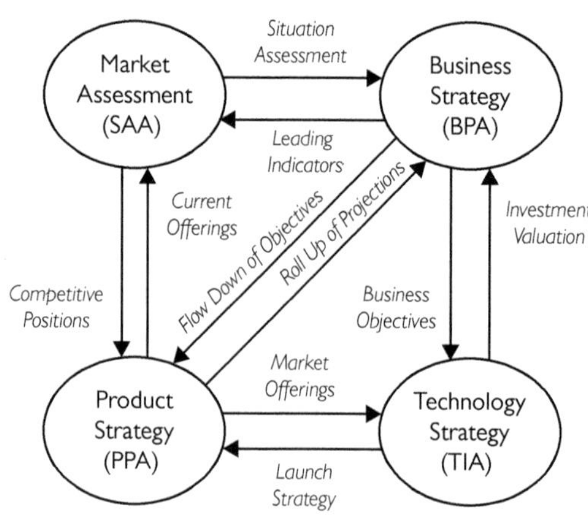

Fig 4.4 The *Advisor Series* of planning tools.

We envisioned the flows between the four tools depicted in Figure 4.4 as being automatic, which would be difficult to accomplish seamlessly. Fortunately, we never attempted this, because customers had a better idea. They were all quite facile with Microsoft Excel. They would capture model output in Excel and then paste them into another model, with perhaps a bit of translation.

The reason they preferred this is that the Excel spreadsheet then became the "minutes" of the working session. Their confidence and expertise with Excel seemed to provide our tools with more credibility. The ability to use their preferred representation, that is, a spreadsheet, increased their level of comfort with the whole process depicted in Figure 4.4. This is an important aspect of acceptability, as defined earlier.[1]

Customers provided another important insight. Notice that the last step in Figure 4.2 is "Generate Documentation." Use of this step would automatically create a set of slides for use with Microsoft PowerPoint. This slide set would capture whatever elements of the product plan the users chose, via a dashboard designed for this. While customers asked for this function within the *Product Planning Advisor*, they seldom used it.

A marketing executive at one of our clients explained this lack of use. He said that he used this function and brought the slide deck to a meeting with his boss for the purpose of requesting budget to proceed. His boss asked how he had created the slides, and the marketing executive opened his laptop and showed him the *Product Planning Advisor*. His boss became intrigued, played around with various assumptions, considered alternative scenarios, and then approved the budget. From then on, the executive always presented the tool itself, never the slides.

We have experienced this phenomenon repeatedly. When participants in *Product Planning Advisor* or *Technology Investment Advisor* sessions can take the controls and explore possible futures, their buy in soars. They also become a rich source of ideas for improving models and visualizations. I think they were also tired of looking at PowerPoint slides.

Evaluation

We conducted a study involving 100 planning teams and over 2,000 participants using one or more of the tools from the *Advisor Series* (Rouse, 1998).

[1] We recently created a new version of the *Technology Investment Advisor* entirely in Microsoft Excel, making learning how to use the tool much easier.

Workshop participants were asked what they sought from computer-based tools for planning and design. Here is a summary of their responses:

- They wanted a clear and straightforward process to guide their decisions and discussions, with a clear mandate to depart from this process whenever they choose.
- They wanted capture of information compiled, decisions made, and linkages between these inputs and outputs so that they could communicate and justify their decisions, as well as reconstruct decision processes.
- They wanted computer-aided facilitation of group processes via management of the nominal decision-making process using computer-based tools and large screen displays.
- They wanted tools that would digest the information that they input, see patterns or trends in it, and then provide advice or guidance that the group felt they would not have thought of without the tools.

Models and simulations can provide the "engine" that drives these capabilities, but greater advisory capabilities are needed to fully satisfy all these objectives. A good example comes from the networked version of the *Product Planning Advisor*, a version that enabled teams to work remotely and asynchronously. When asked what they liked best about this version, users did not comment on the modeling capabilities; instead, they expressed great appreciation for a feature that was unique to this version: the networked *Product Planning Advisor* kept "minutes" of every user transaction with the tool, including every proposed change and implemented change. Given that the tool knew each user and their issues of interest, we had created a function called "What's Happened Since I Was Last Here?" When users clicked on this option, they were provided with an explanation of how their issues of interest had been addressed while they were away. They found this invaluable.

The obvious conclusion from these lessons learned is that people want an environment that helps them address and solve their problems of interest. Computational models coupled to interactive visualizations provide some of this support. However, facilitation—human or otherwise—and capturing of "minutes" are also crucial elements of this support. People really like it when the support system surprises them with suggestions that, upon careful examination, are really good ideas.

Discussion

I discuss many types of models in subsequent chapters in this book. However, none of these others models have been subject to as much evaluation as the *Advisor Series*. My experiences with these four tools involved hundreds of client engagements with many thousands of executives and senior managers.

We helped compute possible futures in industries ranging from automotive to aerospace, electronics to semiconductors, chemicals to pharmaceuticals, to consumer goods and services, to publishers' offerings. These engagements always addressed a range of "what if?" questions. These questions could be informed by data but not answered, because the futures being considered did not exist.

The teams addressing these questions were multidisciplinary, typically representing marketing, engineering, manufacturing, finance, sales, and customer support. They formulated models of the phenomena relevant to their questions, debated assumptions, framed scenarios, and explored the futures of interest.

In the process, there were many model-based predictions. The teams did not expect that their actual future would conform to these predictions. Instead, they provided glimpses into what *might* happen and the conditions under which these possible futures were likely to emerge. This provided them with various model-based insights into their offerings and markets.

In the process, team members learned about each other's perspectives of the problem being addressed, what variables really mattered to them, and what trade-offs they saw as central. In this way, the models and tools spanned the boundaries among the disciplines. People left these sessions with much richer team mental models, as discussed in Chapter 6.

Conclusions

In this chapter, I have outlined several models. The overarching model is human-centered design, in the sense that markets and competitors are characterized by multistakeholder, multiattribute utility functions, and greater success is predicted for products and services that have higher expected utility values than alternative offerings, including those of competitors, do.

The power of this model comes less from the expected utility values calculated than from the discipline of characterizing stakeholders and their preferences, as well as identifying trade-offs across the central issues of validity, acceptability, and viability. This discipline leads multidisciplinary teams to gain new insights and formulate more creative and competitive offerings.

The specific models formulated in the *Product Planning Advisor* varied enormously in terms of the contexts of markets and competitors. These models were employed to explore the sensitivity of outcomes to assumptions and opportunities for improvement, serving as boundary spanning mechanisms for planning teams. The teams, thereby, enhanced their shared mental models of markets and competitors. Subsequent planning sessions became increasingly efficient and effective.

Colleagues have long suggested that I approached my companies as research projects. The market forces and competitive strategies I have studied were, in effect, the phenomena of interest in the laboratories provided by these companies. These studies resulted in my company learning what was important to customers, and what models and tools would best help them to compute their possible futures. My colleagues were correct in their assessment of my intent. The many books and journal articles produced as a result of the research carried out by my companies provide ample evidence of this.

Key Points

- Human-centered design is a process of considering and balancing the concerns, values, and perceptions of all the stakeholders in a design.
- Attributes of importance to stakeholders, and functions that provide these attributes, can be modeled and explored to plan new products and services.
- Tools that are independent but share representations and data can enable the formulation and exploration of these models.
- Model-based analyses should compare planned offerings to those of competitors, including the likely reactions of competitors to your offerings.
- The status quo is compelling, because the customer already has it, knows how to use it, and it requires little, if any, additional expenditures.

REFERENCES

Hauser, J. R., & Clausing, D. (1988). The house of quality. *Harvard Business Review, May–June*, 63–73.

Keeney, R. L., & Raiffa, H. (1976). *Decisions with Multiple Objectives: Preference and Value Tradeoffs*. New York: Wiley.

Rouse, W. B. (1991). *Design for Success: A Human-Centered Approach to Designing Successful Products and Systems*. New York: Wiley.

Rouse, W. B. (1992). *Strategies for Innovation: Creating Successful Products, Systems, and Organizations*. New York: Wiley.

Rouse, W. B. (1993). *Catalysts for Change: Concepts and Principles for Enabling Innovation*. New York: Wiley.

Rouse, W. B. (1998). Computer support of collaborative planning. *Journal of the American Society for Information Science, 49*(9), 832–839.

Rouse, W. B. (2007). *People and Organizations: Explorations of Human-Centered Design*. New York: Wiley.

Rouse, W. B. (2015). *Modeling and Visualization of Complex Systems and Enterprises: Explorations of Physical, Human, Economic, and Social Phenomena*. New York: Wiley.

Rouse, W. B., & Howard, C. W. (1993). Software tools for supporting planning. Industrial Engineering, *25*(6), 51–53.

CHAPTER 5

Technology Adoption

Using the methods and tools in Chapter 4, you created highly competitive products and services. Your revenue and profits are growing nicely. Now what? Can you rest on your laurels? Henry Ford thought he could.

Ford sold the same black car—the Model T—for almost 20 years (1908–27), continually lowering prices as production efficiencies increased. Alfred Sloan at General Motors countered with annual styling changes, as well as a pricing structure such that Chevrolet, Pontiac, Oldsmobile, Buick, and Cadillac did not compete with each other. These concepts, coupled with Ford's resistance to the change, propelled General Motors (GM) to industry-sales leadership by the early 1930s.

Ford's story is not atypical. Kodak, Polaroid, Xerox, Motorola, and Nokia left their potential innovations on the shelf to milk their existing competitive advantages. They delayed introducing new market offerings in order to avoid cannibalizing their existing offerings, for example, film, paper copiers, analog cell phones, and low-end cell phones.

They wanted to milk their cash cows as long as possible. They invested in getting better and better at delivering existing offerings, while their customers wanted less and less of these offerings. They thought, if only implicitly, that business as usual would prevail. It never does. Now these companies are shadows of their former selves.

Achieving enterprise objectives requires increasing the efficiency of existing processes (as Ford did) and/or launching new solutions, which require new processes and/or technologies (as Sloan did). In other words, you need to address the balance between investing in getting better at what you are already doing versus investing in doing new things.

Computing Possible Futures. William B. Rouse. Oxford University Press (2019). © William B. Rouse 2019.
DOI: 10.1093/oso/9780198846420.001.0001

I discuss this balance at length in my undergraduate course on engineering economics. Each year, 500 juniors across all engineering majors hear that investments in processes and technologies should be balanced across what you are already doing versus doing new things. The economic value of potential cash flow streams in these two arenas should inform decisions regarding appropriate levels of investments in these processes and technologies.

The series of initiatives needed to yield these cash flow streams can determine what processes and technologies are needed and the costs of providing them. New product and service offerings often require new technologies to be competitive. This, in turn, requires upstream R & D to yield downstream technology innovations. This upstream R & D can be seen as providing "technology options" to the enterprise's business units. The business units can later exercise one or more of these options if it makes competitive sense at the time. R & D budgets determined in this way are, in effect, "backcasted" from market aspirations.

In this chapter, I discuss how to attach value to technology options. I address the question "What are they worth?" This can be contrasted with the question "What do they cost?" Worth usually has to greatly exceed cost to garner investments. The latter half of this chapter summarizes how we used this approach to enable many well-known technology companies to make major investments in creating technology options that led them to uncover billions of dollars of worth (or value) beyond the costs on these investments.

Options-Based Thinking

Let's begin with the overall line of reasoning of options-based thinking. First, technology and process investments create contingent opportunities for later solution investments. They are contingent in the sense that the options created may not be exercised. Capabilities created upstream with initial investments may not be viable downstream in the marketplace, due to competing technologies, processes, and economic conditions. Thus, these capabilities are options that provide rights to invest but not requirements to invest.

Quite often, people will object to this idea. Why would you invest in capabilities that you may not use? My response to this question is, "Does anyone in the room regret that you did not use your life insurance last year?" Options provide insurance against contingent business needs. These needs may not emerge.

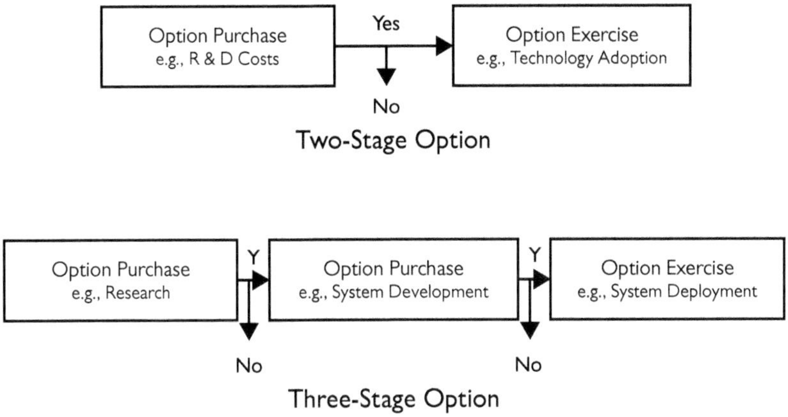

Fig 5.1 Examples of technology options.

If they emerge, the capabilities created may no longer be competitive. Hence, it is good to have a portfolio of options.

Figure 5.1 illustrates two forms of technology options. For the two-stage option, one invests in R & D to create technology-based capabilities. At a later date, one decides whether it makes sense to adopt the technology in a product or service offering from one or more business unit. Adoption amounts to exercising the technology option.

The three-stage option starts with investing in research, for example, chemistry or physics research. If this research is successful, the options created could be exercised by investing in development. If development proves successful and market conditions warrant it, this second option could be exercised by deploying the capability in business units' product and service offerings.

How much should one be willing to invest to buy a technology option? The answer obviously depends on the likely cash flows from later investments to exercise the option. There is usually significant uncertainty regarding these cash flows. Spreadsheet projections often have a single number for each year, but these projections should actually be probability distributions, to reflect the "volatility" of likely cash flows. These uncertainties are often substantial; this is one of the reasons options are attractive.

Uncertainties usually decrease in time due to R & D results emerging, market conditions evolving, and competitors' intentions and capabilities becoming clearer. Thus, delaying decisions to invest in exercising options—rather than deciding now—can be of substantial value. It may be attractive to "purchase" options that can later be exercised (or not).

In general, option values increase with projected cash flows and time, for the reasons just outlined. Option values also increase with uncertainty, which may seem counterintuitive. Many aspects of management are concerned with decreasing and controlling uncertainties. Why would one place higher value on greater uncertainty?

First, keep in mind that the option value of the future of interest increases with the uncertainty associated with this future. Thus, having an option on this future is attractive, in contrast to simply fully investing in this future now. Having highly volatile cash flows in the future means that there is some likelihood of negative cash flows. If one owns an option on this future and later determines that negative cash flow has emerged, one does not exercise this option and invests no further in this future. One the other hand, if positive cash flow has emerged—the option is "in the money"—then greater uncertainty implies a greater likelihood of higher positive cash flows.

The idea of seeking uncertainty does not make sense to many executives. Wouldn't it be better if there were no uncertainty? If this happened, all providers would know exactly what products and services customers wanted. Hence, they would all provide the same offerings, rendering them commodities. Economics tells us that in commodity markets, profit margins asymptotically approach zero as providers continually reduce prices to gain market share. Eventually, the only reason to be in these businesses is because one is already there. Such markets do not make sense for new entrants.

When I served two terms on the U.S. Air Force Scientific Advisory Board, we undertook a study of science and technology investment strategies. The process included interviews of many chief technology officers (CTOs) of top companies. The CTO of DuPont reported that they invested, on average, in 300 R & D projects to bring two ideas to market. Someone asked why he did not pick those two in the first place. He responded by saying, "Our market is inherently laced with uncertainties. We like it that way because we are better at managing uncertainty than our competitors. The reward is high profits on our proprietary products."

Real Options

Real options differ from financial options in that they involve investments in tangible assets rather than financial instruments. Tangible assets include technologies, processes, and even companies, as later examples illustrate.

A "call" option provides the right, but not the responsibility, to buy an asset, with uncertain future value, at a stipulated price before a specified time (American option) or at a specified time (European option). A "put" option, in contrast, provides the right, but not the responsibility, to sell an asset at a stipulated price. We are interested in the value of call options.

The option value equals the expected value of the asset at maturity (EVA), given that the EVA is greater than the option exercise price (OEP) minus the OEP multiplied by the probability that the EVA is greater than the OEP. Net option value (NOV) equals the option value calculated in this manner minus the discounted option purchase price (OPP).

If the option is "in the money," that is, EVA is greater than OEP, then the option will be exercised. If the EVA is less than or equal to the OEP, the option will not be exercised, and no more money will be invested. In this way, the option hedges the downside risk of the EVA being less than or equal to the OEP.

Fisher Black and Myron Scholes (1973) and Robert Merton (1973) formulated approaches to determining the option value, assuming a portfolio of a security S and a risk-free bond B. The evolution of the value of S is represented by a first-order differential equation with added volatility, characterized by σ. The value of the bond evolves deterministically with a fixed interest rate, r. The purpose of the bond is to generate the OEP.

This representation leads to a partial differential equation—the Black–Scholes equation—expressing the option value as a function of S and time. Solution of this equation for a European call option results in the option value being expressed in a simple algebraic equation. A European call option can only be exercised on its expiration date.

The Black–Scholes algebraic solution provided a core element of the Technology Investment Advisor discussed in the next section. Scholes and Merton were awarded the 1997 Nobel Prize in Economics for this work on option pricing. Fisher Black died in 1995 and therefore was not eligible to be included in the award.

Technology Investment Advisor

The Technology Investment Advisor provides a means to answering the worth question (Rouse, et al., 2000; Rouse, 2007). Our initial exploration into using a real-option approach to the valuation of technology investments was for the U.S. Air Force. We quickly ran into difficulties, as the requisite data were unavailable. Motorola, another client at the time, soon invested in sustaining this exploration.

They posed the question as follows. They were then investing about 12 % of revenues in R & D. This was based in part on the behaviors of their competitors, but they admitted that they had no idea whether this percentage should be higher of lower. In a two-day-long discussion of how to approach this issue, I suggested that we first discuss why it is not zero.

The response was straightforward. With zero investment, their business units would not have the technology capabilities that provide competitive advantages. What advantages did they need? Performance, cost, and time to market were among their answers. I suggested that we think about how to provide technology options for these capabilities. This was serendipitous, as Motorola's CTO had recently become interested in real options. We were off and running.

The Technology Investment Advisor emerged to function as follows. Production S-curves are used to project the number of units sold over time; the curves plateau and then decline when new generations of products are introduced and begin to ascend their own S-curves. Production costs follow classical learning curves; for example, every doubling of the number of units produced results in unit production costs of 60–80 % of the earlier costs. Making it down the learning curve, while maintaining prices, is key to steadily increasing profits, unless all the competitors in the market are experiencing the same learning curves.

Projections of units produced, costs per unit, and prices per unit enable predictions of an enterprise's future financial statements, based on the upstream options in which the enterprise has invested. The concern is with the option values of these possible futures. Typically, the option value has to be substantially higher than the OPP, that is, the NOV has to be several multiples of OPP for investments commitments to be made.

The Technology Investment Advisor was a proprietary software package with a dashboard allowing manipulations of various parameters, probability distributions of these parameters, and, hence, projected probability distributions of NOV, and numerous visualizations of projections. A few years ago, clients asked for an updated version of the Technology Investment Advisor. We found that the capabilities of Microsoft Excel had advanced sufficiently that the latest version of the Technology Investment Advisor could be fully accommodated in Excel. That is the version we have employed for our latest engagements.

The Technology Investment Advisor is often used in conjunction with the *Product Planning Advisor*, as shown in Figure 5.2. The *Product Planning Advisor*

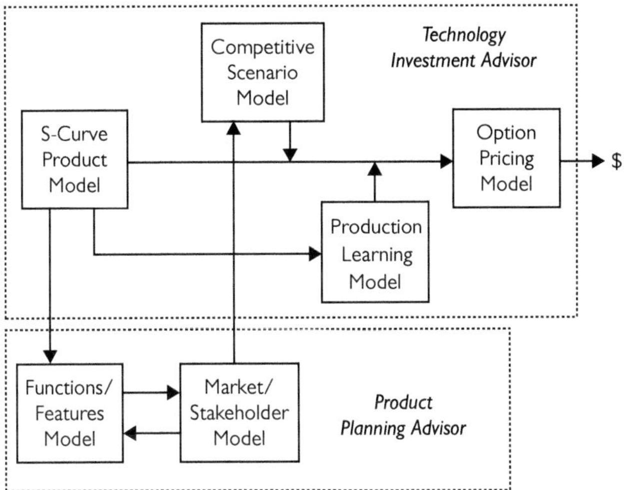

Fig 5.2 Technology options enable product offerings.

is used to determine the desired functions and features of a product of service, as well as relative competitive advantages of alternative offerings. The Technology Investment Advisor is then used to project the financial attributes of the technologies needed to enable these offerings. The next section summarizes numerous case studies of this process.

Considering the capabilities depicted in Figure 5.2 in terms of the representational paradigms discussed in Chapter 2, three are employed:

- dynamic systems theory: the dynamics of market penetration and decline are represented by generations of S-curve models; dynamic learning curves are used to represent impacts of production levels on production costs;
- decision theory: rule-based competitive scenario models are used to represent market preferences, as influenced by the *Product Planning Advisor* models based on utility theory and network representations;
- finance theory: option pricing models are used to estimate the values of technology options; discounted cash flow models are used to calculate NPVs for all financial variables.

Thus, three of the eight paradigms from Chapter 2 are integrated into this computational model of how technology options enable market offerings.

Case Studies

Over the past 20 years, I have led roughly 30 engagements with clients interested in applying the real-options approach to assessing the worth of technology investments. In this section, I discuss 14 of the engagements that involved Lockheed Martin, 3M, Motorola, Raytheon, and the Singapore Ministry of Defense. The engagements not discussed addressed the consumer-products, forest-products, pharmaceutical, publishing, and shipbuilding industries.

The engagements discussed in this section involved one or more of the following objectives:

- investing in R & D: in many cases, options are bought by investing in R & D to create "technology options" that may, or may not, be executed in the future by business units;
- running the business: in this case, option was bought by continuing to operate an existing line of business, perhaps with modest current profitability but with substantial future opportunities;
- acquiring capacity: this case involves the possibility of acquiring capacity, to support market growth that is uncertain in terms of timing, magnitude, and potential profitability;
- acquiring competitor: this case involves the possibility of acquiring competitors to support growth in current and adjacent markets that is uncertain in terms of timing, magnitude, and potential profitability.

For all four types of objectives, the option purchase involved an investment that was modest compared to the level of investment needed to exercise the option at a later date—if that made sense at that time. If the lion's share of the investment must be made up front, then standard NPV analyses are warranted, rather than NOV analyses.

An overview of 14 case studies gleaned from 30 engagements is shown in Table 5.1. The "Option Purchase" column classifies these case studies in terms of the above four categories. The "Option Exercise" column indicates the activities needed if they later choose to exercise the option. The "Net Option Value" column shows the NOV, which, as defined earlier, is the option value minus the OPP.

Note that the total NOV across the 14 engagements was $4.2 billion. This is the value beyond the purchase prices of the options, which averaged $10–20 million. Consider the second row from the bottom. This case involved an R & D

Table 5.1 ***Results of options-based analyses***

Technology	Option Purchase	Option Exercise	Net Option Value ($ Million)
Aircraft (manufacturing)	R & D	Deploy improvements	8
Aircraft (unmanned)	R & D	Deploy system	137
Auto radar	Run business	Expand offerings	133
Batteries (lithium ion)	R & D	License technology	215
Batteries (lithium polymer)	R & D	Acquire capacity	552
Fuel cell components	R & D	Initiate offering	471
Microsatellites	R & D	Deploy system	43
Optical multiplexers	R & D	Expand capacity	488
Optical switches	Run business	Expand offerings	619
Security software	Run business	Add market channels	267
Semiconductors (amplifiers)	Invest in capacity	Expand offerings	431
Semiconductors (graphics)	R & D	Initiate offering	99
Semiconductors (memory)	R & D	Initiate offering	546
Wireless LAN	Run business	R & D	191

investment by Motorola in magnetoresistive random access memory, where data is not stored as electric charge but in magnetic storage.

The research team developing this technology was requesting $20 million. As shown in Table 5.1, the NOV was $546 million. After carefully listening to a presentation of the Technology Investment Advisor basis of this estimate, the Motorola CEO was sufficiently impressed to commit $40 million, with the request that the additional funds be used to reduce risk and accelerate creating this technology option for their semiconductor business.

The success of this technology contributed to the formation of Freescale Semiconductor, Inc., which was spun off from Motorola in 2004. In 2015, NXP Semiconductors completed its acquisition of Freescale for about $11.8 billion in cash and stock. Including the assumption of Freescale's debt, the purchase price was about $16.7 billion. The option-based approach contributed significantly to launching this success.

The second row from the top of Table 5.1 involved R & D on a large, unmanned aircraft for surveillance and reconnaissance of the island nation of Singapore. Upon being presented the Technology Investment Advisor analysis of this investment, with an NOV of $137 million, the Minister of Defense immediately approved the investment. Like the Motorola CEO, the minister wanted to also know what it would cost to reduce risk and accelerate creating this technology option for their defense forces.

These two examples emphasize how the Technology Investment Advisor is used. Decision-makers do not simply compare NOV to OPP and decide whether to invest. They like to explore the sensitivity of the projections to assumptions, for example, discount rate, inflation, and rates of growth of market share. They often ask teams presenting their Technology Investment Advisor analyses how they will counter technology and market risks.

Technology Adoption in Automobiles

One of the case studies in Table 5.1 involved automobiles; namely, it concerned auto radar. We have continued our pursuits of applications of options-based thinking in the automobile industry. Technology adoption occurs in the context of overall processes of conceptualizing, planning, designing, developing, and selling new vehicles. This section briefly considers this overall context.

Several years ago, we conducted a study for GM of the best 10 and worst 10 cars, industry wide, over the past 50 years (Hanawalt & Rouse, 2010). GM's executives and managers chose the 20 cars listed in Table 5.2. The question was, "How were the decisions made to put these cars into the market, and what distinguished successes from failures?"

Answering this question required quite a bit of sleuthing. Beyond looking at production and sales numbers, we found a wealth of interviews of automotive executives and chief engineers, as well as published reviews of these cars.

Central success factors included

- having correct assumptions about the target segment of a new entry in the automotive market several years in the future;
- employing a vehicle development process that could reliably and efficiently yield the vehicle envisioned; and
- having designers and developers who were able to adapt to changing economic conditions in the marketplace.

Table 5.2 ***Ten product successes and 10 product failures***

Product Successes	Product Failures
1955 Chevrolet	1958 Edsel
1964 Pontiac GTO	1960 Chevrolet Corvair
1964 Ford Mustang	1971 Chevrolet Vega
1974 Volkswagen Rabbit	1971 Ford Pinto
1984 Chrysler Minivan	1975 AMC Pacer
1986 Ford Taurus	1981 DeLorean DMC-12
1998 Lincoln Navigator	1982 Cadillac Cimarron
1998 Toyota Prius	1988 GM10 Program
2001 Chrysler PT Cruiser	1990 Saturn
2003 Cadillac CTS	2001 Pontiac Aztec

Leadership also played a role, especially when capricious decisions were made, which tended to throw the evolution of the vehicle off track. The Pontiac Aztec is a good example, where a last-minute decision was made to put the vehicle on a minivan chassis, without redesigning the body due to this change. The result was "the ugliest car on the road," according to various pundits, including *Car Talk*. One GM executive commented to me, "It is really not a bad car when you are driving it." I responded, "The best thing about driving it is that you then cannot see it."

Another management fiasco was the Cadillac Cimarron, which was basically a Chevrolet Cavalier with a Cadillac badge, selling for $10,000 more than the Chevrolet. GM apparently thought that customers would only see the badge. Most auto publications quickly labeled the Cimarron a complete flop. Sales were far below expectations.

There are two bottom-line lessons learned. First, cars should be targeted at markets, as they will be when the cars hit the market. Edsel was targeted at an upscale market but appeared during the 1957 recession. Second, the product development process should yield the desired car, with sufficient discipline to minimize capricious management decisions.

The results of this study caused us to wonder why cars are removed from the market. We studied 12 cars that were withdrawn from the market in the 1930, the 1960s, and the 2000s (see Table 5.3; Liu, Rouse, & Yu, 2015). Using the same

Table 5.3 ***Twelve vehicles withdrawn from the market***

Year	Vehicle
1937	Duesenberg
1937	Cord
1938	Pierce-Arrow
1940	LaSalle
1958	Packard
1960	DeSoto
1966	Studebaker
1969	Rambler
2001	Plymouth
2004	Oldsmobile
2010	Pontiac
2010	Mercury

methodology as used for the earlier study, we assessed influences at the levels of economy, market, company, and vehicle.

Factors that had pervasive influence included economic crises, market declines, poor management, external competitors, and company financial problems. The vehicle itself, particularly in terms of quality, was only a pervasive issue in the 2000s, when global competitors provided much higher-quality vehicles.

These failures reflected, to a great extent, inabilities to balance the tension between differentiated offerings and economies of scale or market demands. This was quite clear among U.S. auto brands during the 2000s. GM, for example, sought economies of scale by making the Chevrolet, Pontiac, Oldsmobile, and Buick brands virtually identical except for their badges. Brand loyalty progressively became meaningless.

What is clear from these two studies is that the success of a vehicle depends on much more than just the vehicle. Success is influenced by those technologies that are adopted and subsequently embraced by the marketplace. However, poor management and bad financial decisions can undermine these advantages. Economic crises and wars can further worsen the state of enterprises.

The two studies just summarized were retrospective, case-based assessments focused on explaining past decisions, using a wide variety of sources of numeric and textual information. The next two studies, supported by Accenture, were prospective projections of likely impacts of new technologies on the automotive marketplace.

These studies addressed adoption of new powertrain technologies (Liu & Rouse, 2015; Liu, Rouse, & Hanawalt, 2017) and driverless car technologies (Liu, Rouse, & Belanger, 2019). Both studies involved developing models based on dynamic systems theory that included elements for manufacturers, consumers, and government. The second study added the insurance and auto finance components of the automotive ecosystem.

The study of new powertrain technologies addressed adoption scenarios in terms of government rebates, manufacturers' willingness to invest, and consumers' purchasing preferences. Both economic and environmental impacts were projected. Long-term impacts were projected to come primarily from product familiarity, consumer preferences, and technology competitiveness, rather than transitory government rebates. There is substantial potential for electric vehicles to provide environmental benefits. However, these benefits will not be realized as long as vehicles are recharged via electricity generated by coal-fired power plants.

The second study focused on adoption of autonomous vehicles as affected by consumer choice and technology acceptance. We considered both direct consumer purchases of vehicles and consumers' use of car services. Vehicle sales were projected, as were insurance premiums collected.

Most states regulate premiums to equal the costs of claims. Hence, the reduced accident rates of autonomous vehicles resulted in dramatic declines of premiums collected over the coming decades. While not reported in this paper, auto finance also sees substantial declines in revenues, as the typical four loans per vehicle (once when new; three when used) is undermined by high-use vehicles being recycled after their first owners log hundreds of thousands of miles per vehicle.

These two studies showed that the impacts of these technologies range far beyond the vehicle and auto manufacturers. The whole transportation ecosystem is likely to be affected. When we presented these results to GM, they commented that they had been assuming that the primary impacts of change would be on them. It was clear from our results that other players would be strongly impacted in ways that GM had not anticipated.

This chapter has focused on estimating the economic value of investing in technologies that might enable new market offerings. This section has addressed an example of the context within which such investment decisions are made.

There is much more involved than just estimating NOVs. After one of the studies summarized in Table 5.1, the CEO of the subsidiary company involved was quite enthusiastic about the resulting projections. He liked the new numbers, but then, with a little reflection, commented, "You have to count the numbers right, but the numbers are not all that counts."

Organizational Implications

How does options-based thinking change the way an organization operates? Investment decision criteria change, as elaborated below. Beyond this direct change, there are broader organizational implications of adopting options-based thinking that are outlined later in this section.

An enterprise simulation, R & D World, was used to assess the merits of alternative decision criteria for R & D investments (Bodner & Rouse, 2007). This case study focused on R & D investments by a large forest-products company. This simulation was used to compare criteria of NPV, NOV, and this company's other stage-gate criteria, for example, strategic fit.

The R & D World simulation was unveiled as a meeting of CTOs of leading forest-products companies. Only one of the CTOs knew that the company being simulated was his. Each CTO was asked to propose what he or she would use for decision criteria. The 15 proposals were simulated for 10 replications of 25-year periods. The resulting average profits ranged from a loss of $250 million to profits of $1.55 billion.

The CTOs were shocked. One commented, "I had no idea these criteria could make so much difference." A key distinction emerged. Emphasis on NPV preserves the R & D budget. Indeed, often the R & D budget was not fully expended, due to a lack of positive NPV proposals.

In contrast, emphasis on NOV results in all, or nearly all, of the budget being expended. More projects are funded—modestly at first, as they are options. The company has more options that may (or may not) be exercised. Rather than preserving the R & D budget, emphasis on NOV maximizes earnings per share. It seems reasonable to assume that this is the primary purpose of the R & D budget.

Peter Boer (1999), former CTO of chemical conglomerate W. R. Grace, has creatively addressed the question of how best to reconcile NPV and NOV. He proposed a new metric, "strategic value," which is the sum of NPV and NOV. NPV analyses should focus on current lines of business, while NOV should address aspirational lines of business. This directly addresses the question of investing in getting better at what you are currently doing versus investing in doing new things.

A good example is "Wireless LAN," in the bottom row of Table 5.1. The CFO of the client company was considering shedding this provider of networking services to commercial customers. The business plan showed an NPV of $3 million, pretty small potatoes for a $20-billion-dollar company. We assessed the possibility of this company addressing the home networking market, which had not yet emerged.

The option, as indicated in Table 5.1, was to simply keep running this small company while creating plans and capabilities to enter the home networking market. The NOV for this plan was $191 million. I was scheduled to present the results to the CFO. At the last minute, the vice president with whom I had been working told me that my presentation could be only slide long. "Pick your best one," he said.

I entered the CFO's office and laid one page, a bar graph, on his conference table. One bar was $3 million high and labeled NPV. The other bar was $191 million high and labeled NOV. In typical CFO fashion, he asked, "What do I have to do to get the big bar?" I told him that he just had to keep the small company running until it was clear the home market was emerging, which would probably be after about three years. Then, if warranted, he could exercise his option for $12 million. That is what he did.

Options-based thinking provides a framework for creating value-centered organizations (Rouse & Boff, 2004). The first concern is *characterizing value*. As this chapter has elaborated, value is created in R & D organizations by providing "technology options" for meeting the contingent needs of the enterprise. In this way, R & D organizations provide a primary means for enterprises to manage uncertainty by generating options for addressing contingent needs. A central challenge for R & D organizations is to create a portfolio of viable options. Whether or not options are exercised is an enterprise challenge.

The next concern is *assessing value*. Value streams, or value networks, provide a means for representing value flow and assessing the value of options created. Valuation of R & D investments can be addressed by assessing the value of the options created in the value network. The options-based approach outlined earlier in this chapter provides the means for this assessment.

A third and crucial concern is *managing value*. Decision-making processes—governance—are central to managing the flow of value. Specifically, if NOV is communicated as a key metric, then this metric should affect decisions. If other factors dominate, for example, who knows whom, then emphases on NOV will be quickly and cynically dismissed.

Organizational structure affects value flow, with significant differences between hierarchical versus heterarchical structures. In two studies of R & D organizations, one in industry and one in government, we found that having to appeal to the hierarchy for permission or resources resulted in delays of execution for months and sometimes years.

Individual and team affiliations and identities affect value flow; dovetailing processes with disciplines is essential. People need to be affiliated with value streams *and* their disciplinary base. Without the former, recognitions of contributions are likely diminished. Without the latter, disciplinary expertise can wither.

Champions play important, yet subtle, roles in value flow. Supporting champions is necessary but not sufficient for success. In one study, we encountered a situation where deployment of R & D outcomes only occurred when champions found ways to circumvent processes intended to help them. This is not a sustainable approach to value.

Incentives and rewards affect value flow. Aligning these systems with value maximization is critical. If the incentive-and-reward system remains aligned to outdated value propositions, people in the organization will continue to march to the old drummer. A good example is academia, where multidisciplinary research is extolled but incentives and rewards remain tightly tied to individual accomplishments.

Successful technology adoption depends on good ideas, adequate resources, and hard work, but that is not enough. Value needs to be characterized and assessed appropriately. Value needs to be managed to align behavioral and social phenomena with the value proposition being pursued. Misalignment can undermine ideas, resources, and work.

Key Points

- You need to address the balance between investing in getting better at what you are already doing versus investing in doing new things.
- There are numerous high-profile examples of companies unsuccessfully addressing this balance, leading to their eventual demise.
- Technology and process investments create contingent opportunities for later solutions; they are contingent, in the sense that they may not emerge.
- Many markets are inherently laced with uncertainties. High profits can result if you are better at managing uncertainty than your competitors.

- Strategic value is the sum of the NPV of current lines of business, and the NOV of aspirational lines of business.
- Options-based thinking can provide a framework for value-centered organizations by characterizing value, assessing value, and managing value.

REFERENCES

Black, F., & Scholes, M. (1973). The pricing of options and corporate liabilities. *Journal of Political Economy, 81*(3), 637–659.

Bodner, D., & Rouse, W. B. (2007). Understanding R&D value creation with organizational simulation. *Systems Engineering, 10*(1), 64–82.

Boer, F. P. (1999). *The valuation of technology: Business and financial issues in R&D*. New York: Wiley.

Hanawalt, E. S., & Rouse, W. B. (2010). Car wars: Factors underlying the success or failure of new car programs. *Systems Engineering, 13*(4), 389–404.

Liu, C., & Rouse, W. B. (2015, April 1). Ten years of top ten tech cars: An analysis. *IEEE Spectrum*. Retrieved from http://spectrum.ieee.org/cars-that-think/transportation/advanced-cars/ten-years-of-top-ten-tech-cars-an-analysis.

Liu, C., Rouse, W. B., & Belanger, D. (2019). Understanding risks and opportunities of autonomous vehicle technology adoption through systems dynamic scenario modeling: The American insurance industry. *IEEE Systems Journal*, in press.

Liu, C., Rouse, W.B., & Hanawalt, E. (2018). Adoption of powertrain technologies in automobiles: A system dynamics model of technology diffusion in the American market. *IEEE Transactions on Vehicular Technology, 67*(7), 5621–5634.

Liu, C., Rouse, W. B., & Yu, X. (2015). When transformation fails: Twelve case studies in the automobile industry. *Journal of Enterprise Transformation, 5*(2), 71–112.

Merton, R. C. (1973). Theory of rational option pricing. *Bell Journal of Economics and Management Science, 4*(1), 141–183.

Rouse, W. B. (2007). *People and organizations: Explorations of human-centered design*. New York: Wiley.

Rouse, W. B., & Boff, K. R. (2004). Value-centered R&D organizations: Ten principles for characterizing, assessing, and managing value. *Systems Engineering, 7*(2), 167–185.

Rouse, W. B., Howard, C. W., Carns, W. E., & Prendergast, E. J. (2000). Technology investment advisor: An options-based approach to technology strategy. *Information Knowledge Systems Management, 2*(1), 63–81.

CHAPTER 6

System Failures

In Chapter 4, I considered markets and competitors and discussed a model-based approach to planning new product and service offerings. Chapter 5 addressed investment in technologies to enhance these offerings. This chapter focuses on the operations and maintenance of these offerings once deployed. In particular, we address dealing with system failures.

Various pundits would have us believe that AI and autonomous systems will operate flawlessly, addressing and remediating failures without human intervention. However, systems fail for a variety of reasons, not all of which can be anticipated. We usually expect the humans in these systems to detect, diagnose, and compensate for these failures. Can we predict their behaviors (what they do) and performance (how well they do) in these tasks? To address this question, let's consider a couple of examples.

Eastern Air Lines (EAL) Flight 401 crashed into the Florida Everglades at 11:42 p.m. on December 29, 1972, causing 101 fatalities. All 3 pilots of this Lockheed L-1011, 2 of its 10 flight attendants, and 96 of its 163 passengers died; 75 passengers and crew survived.

The crash occurred while the entire flight crew was preoccupied with a possibly burned-out landing-gear indicator light. They were concerned with whether or not the landing gear was down. They failed to notice that they had inadvertently disconnected the autopilot. As a result, the L-1011 gradually lost altitude and crashed into the Everglades.

We replicated this scenario in a flight simulator, in this case using two-person flight crews. Three of the four crews encountered the exact same difficulty. The possibility of a failed light bulb captured their attention, and they

Computing Possible Futures. William B. Rouse. Oxford University Press (2019). © William B. Rouse 2019.
DOI: 10.1093/oso/9780198846420.001.0001

were distracted from controlling the aircraft. The simulator could not crash, but they encountered analogous dangers.

These crews needed to simultaneously control the system and diagnose the failure. How good are people at doing this? Can we computationally model such situations and gain useful insights? How can we better understand people being captured by the problem-solving and not being vigilant to the control task?

The second example involves the Aegis cruiser USS *Vincennes* shooting down an Iranian passenger airliner in the Persian Gulf on July 3, 1988 (Rogers, et al., 1992). The Aegis weapon system was first commissioned in 1983 with the USS *Ticonderoga*. This system was developed to counter the serious air and missile threat that adversaries posed to U.S. carrier battle groups and other task forces.

The *Vincennes* incident prompted a congressional inquiry. Subsequently, the Office of Naval Research established a research program to study the potential behavioral and social factors underlying this incident. The TADMUS Program was named for its focus—tactical decision-making under stress. I was the principal investigator for one subcontractor's efforts in this program.

We studied the crews' training. We reached two conclusions. First, crews were "captured" by their training. Every scenario involved 25 folks facing Armageddon. The Iranian commercial airliner was far from Armageddon, but if that were the only scenario you have ever seen, what would you expect?

The second conclusion concerned teams' mental models, a topic I explore in great depth later in this chapter. Put simply, at this point, teams did not understand what each other was doing and what each team member expected of other team members. If the wide receiver does not expect that the quarterback will throw him the ball, team effectiveness can be substantially undermined, and the game may be lost.

We would like to predict human behaviors (what they do) and human performance (how well they do) in situations such as those described above. We want to understand possible futures, often as a function of how people are trained as well as how they are aided. Training affects people's potential to perform, while aiding directly augments their performance. I discuss training and aiding later in this chapter, as well as in Chapter 8.

The human behaviors of interest include humans' formation of perceptions, expectations, and intentions; humans' problem-solving, planning, decision-making, and control; and how behaviors vary in individual versus team versus group settings. We want to model these behaviors to predict how humans are likely to respond to situations requiring particular behaviors.

Human Behavior and Performance

As discussed in Chapter 2, there are many interpretations of the word "model." There are fashion models, this year's car models, and toy models for children and, occasionally, adults. The meaning in this chapter is much more focused. A model is a mechanism for predicting human behavior and performance for specific tasks in particular contexts.

One approach to such modeling involves using empirical data that has been previously collected and that is applicable to the tasks and contexts of interest. Thus, one might, for example, have data on visual acuity as a function of character font and size, as well as ambient lighting. This might take the form of several equations relating the probability of reading errors to character size and lighting level for each font type. One would assume that these relationships would hold for the new task and context; then, by interpolating and carefully extrapolating, it should be possible to use this data to predict reading errors in the new situation.

This approach to modeling is termed "descriptive," in the sense that one describes past behavior and performance, assumes these data apply to the current tasks and context, and interpolates or extrapolates to make predictions. In effect, one is saying that the past data describes how humans performed and that such descriptions provide good predictions of how humans will perform in the new situation.

This works well for aspects of behaviors and performance that are not too context sensitive. Human abilities to see displays, reach controls, and exert force are good examples. This approach is much less applicable to phenomena such as those associated with the earlier examples. The limitation of the descriptive approach is the difficulty of compiling the many task- and context-specific empirical relationships in advance of encountering new design issues and problems.

The "prescriptive" approach focuses on how humans should behave and perform. This begs the obvious question of whether people can conform to such prescriptions. This reality led to the notion of "constrained optimality." Succinctly, it is assumed that people will do their best to achieve task objectives within their constraints, such as limited visual acuity, reaction-time delays, and neuromotor lags (Rouse, 1980, 2007, 2015).

Predicted behaviors and performance are calculated as the optimal behavior and performance subject to the constraints limiting the humans involved. Typically, if these predictions do not compare favorably with subsequent empirical measurements of behaviors and performance, one or more constraints have been missed.

Determining the optimal solution for any particular task requires assumptions beyond the likely constraints on human behavior and performance. Many tasks require that humans understand the objectives or desired outcomes and inputs to accomplish these outcomes, as well as any intervening processes. For example, drivers need to understand the dynamics of their automobiles, pilots need to understand the dynamics of their aircraft, and process-plant operators need to understand the dynamics of their factories. Humans also need to understand any trade-offs between, for example, accuracy of performance and energy expended, human or otherwise, to achieve performance.

This understanding is often characterized as humans' "mental models" of their tasks and context. To calculate the optimal control of a system, the optimal detection of failures, or the optimal diagnoses of failures, assumptions are needed regarding humans' mental models. If we assume well-trained humans who agree with and understand task objectives, we can usually argue that their mental models are accurate, that is, they reflect the actual physical dynamics of the vehicle, aircraft, or factory.

For novice humans, this is an unwarranted assumption. For humans that may or may not agree with task objectives, this assumption may also be indefensible. For these and other reasons, models of social phenomena are a greater challenge. However, within the realm of highly trained professionals operating within engineered systems, the prescriptions of constrained optimality have been rather successful.

Example Models

There are many expositions of models of human behavior and performance that can be drawn upon when concerned with human phenomena. There have been two studies by the National Academies (Pew & Mavor, 1998; Zacharias et al., 2008), and numerous books published on this topic (Rasmussen, 1986; Rasmussen, Pejtersen, & Goodstein, 1994 ; Rouse, 1980, 2007, 2015; Sheridan, 1992, 2017; Sheridan & Ferrell, 1974). The catalog is quite rich. A few examples illustrate this richness.

Manual Control

Manual control, in contrast to automatic control, involves a human generating a control action in an attempt to cause the state of a process to assume desirable values. As shown in Figure 6.1, the human is assumed to observe a noisy

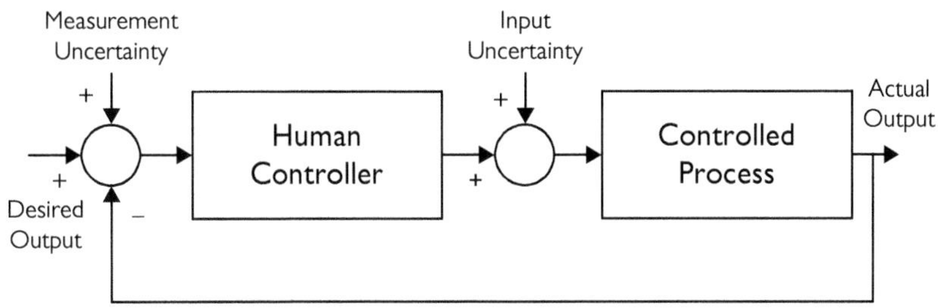

Fig 6.1 Block diagram of manual control.

version of the difference between the desired output and the actual output of the process. The noise is added to reflect the fact that the human is unlikely to observe or measure this difference exactly. The input to the process is also assumed to be noisy due to the neuromotor noise that occurs when the controls are moved.

For the task depicted in Figure 6.1 (repeated from Chapter 1), we can view the human as an error-nulling device. In control theory terms, we can say that the human acts as a servomechanism. Using this servomechanism analogy, we can proceed to analyze the human-process system as if the human were, in fact, a servomechanism.

Manual control has a rich history. Progressively more sophisticated problems of display and control design have been addressed. For example, what if the human were provided with a preview of the future desired output, just as automobile drivers can see the road ahead? What if the human were provided with a display of the predicted actual output to compare to the preview of the future desired output? Intuitively, one would think that these two types of information would improve performance. Manual control models have been developed to predict the extent of improvement, and empirical human-in-the-loop studies have supported these predictions.

In keeping with the notion of constrained optimality, manual control in the highly constrained context of landing an aircraft is close to automatic control, within humans' constraints of not being able to sense variables exactly or execute actions precisely. Interestingly, aircraft are now increasingly flown on automatic control, with the flight crew supervising the automation.

Multitask Decision-Making

The EAL Flight 401 crew failed in their supervisory control of the aircraft. This was due, at least in part, to their need to address multitask decision-making. Figure 6.2 depicts a situation where a human has to perform multiple tasks. For

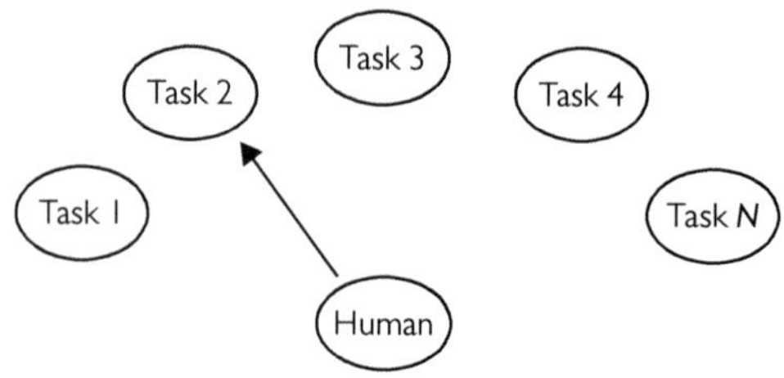

Fig 6.2 Multitask decision-making.

example, a human driving a car has several tasks—lateral control (steering), longitudinal control (accelerating/decelerating), scanning instruments, and so on. Another example is the aircraft pilot whose tasks include lateral control, longitudinal control, communicating with air traffic control, checking and updating radio fixes, monitoring the aircraft's subsystems, and so on. Yet another example is the human who monitors multiple processes in a power plant or a chemical plant.

In most situations, not all of the tasks require attention simultaneously. In other words, tasks "arrive" at different times. Many tasks, such as emergencies, arrive randomly. Some tasks are, of course, scheduled. However, if the schedule is very loose, such as for aircraft arriving at an airport, the flow of arrivals is quite likely to appear random. The human has to decide which arrivals need attention when. High-priority tasks may need immediate attention, while lower-priority tasks can be performed when time is available. The crew of EAL Flight 401 obviously felt that the state of the landing gear was a high priority.

Queuing theory has been used to model human decision-making in multitask situations. Various priority queuing models have been employed and empirically validated for complex multitask environments like flight management. These models would predict that the flight crew would balance supervisory control of the aircraft and solving the problem of the landing gear, perhaps by having one crew member monitor flight performance while the other two crew members addressed the landing-gear issue.

However, they did not expect the flight management task to "arrive," because it was delegated to the automation, at least until someone bumped the controls. It is easy to imagine a similar scenario in driverless cars. However, technology for assessing whether a human operator is paying attention is much more sophisticated now. If you bump the controls and the automation disengages, the car will assess whether you have taken over the controls and clearly warn you if it cannot sense this.

Problem-Solving

The flight crew's problem-solving behaviors involved addressing the state of the landing gear and determining whether the light bulb had failed or the landing gear had, in fact, not firmly engaged. There are models that can help us understand these behaviors.

Early studies of human problem-solving began in the 1930s. Research in the 1960s and 1970s focused on simple, albeit novel, laboratory tasks, for example, the Tower of Hanoi. Allen Newell and Herbert Simon's *Human Problem Solving* (1972) is seen as a classic resulting from such studies. Their rule-based paradigm, which they termed production systems, became a standard in cognitive science.

More recently, greater emphasis has been placed on the study of real problem-solvers performing real-world tasks. Research on the detection and diagnosis of systems failures (Rouse, 2007; Rasmussen & Rouse, 1981) is a good example. Rouse and Spohrer (2018) address situations where AI can be used to augment human problem-solving, in domains ranging from driverless cars, to healthcare, to insurance underwriting.

Typical assumptions associated with models based on problem-solving theory include a specified human mental model of the problem domain and known information utilization behaviors, repertoire of symptom patterns, and troubleshooting rules. The phenomena usually predicted by such models include time until problem solved, steps until problem solved, and frequency of problem-solving errors.

Drawing upon a wide range of sources, I developed a general three-level representation of human problem-solving (Rouse, 1983, 2007). Rasmussen's distinctions among skill-based, rule-based, and knowledge-based behaviors (Rasmussen, 1983), in combination with Newell and Simon's (1972) theory of human problem-solving, led to the conclusion that problem-solving occurs on more than one level—see Table 6.1.

When humans encounter a decision-making or problem-solving situation, they usually have some expectations associated with the context of the situation. They perhaps unconsciously invoke a frame (Minsky, 1975) associated with this situation. Frames describe what one can expect in an archetypal situation, for example, a wedding versus a funeral.

Based on the frame, they then consider available information on the state of the system. If this information maps to a familiar pattern, whether normal or abnormal, it enables them to activate scripts (Schank & Abelson, 1977) that

Table 6.1 ***Problem-solving decisions and responses***

	Decision	State-Oriented Response	Structure-Oriented Response
Recognition and classification	Frame available?	Invoke frame	Use analogy and/or basic principles
Planning	Script available?	Invoke script	Formulate plan
Execution and monitoring	Pattern familiar?	Apply appropriate S-rules	Apply appropriate T-rules

enable them to act, perhaps immediately, via recognition-primed symptomatic rules (S-rules) that guide their behaviors (Klein, 2004).

If the observed pattern of state information does not map to a familiar pattern, humans must resort to conscious problem-solving and planning (Johannsen & Rouse, 1983), perhaps via analogies or even basic principles. Based on the structure of the problem, which typically involves much more than solely observed state variables, they formulate a plan of action and then execute the plan via structural topographic rules (T-rules). As this process proceeds, they may encounter familiar patterns at a deeper level of the problem and revert to relevant S-rules.

This framework has important implications for multilevel modeling of complex systems laced with behavioral and social phenomena. Succinctly, it may not make sense to represent human behavior and performance for any particular task by using only one type of model. Scripted behaviors may be reasonable for familiar and frequent instances of these tasks. However, for unfamiliar and/or infrequent instances of these tasks, a more robust representation is likely to be needed.

We performed extensive studies of human diagnosis of systems failures in aircraft, automobile and ship-propulsion systems; power and process plants; and communications systems. A series of problem-solving models was developed and evaluated relative to actual troubleshooting behaviors and performance (Rouse, 1983, 2007). The last and most comprehensive model was a fuzzy rule-based model.

Consider the above notions of S-rules and T-rules. S-rules are very powerful. Patterns of cues provide prompt immediate, and occasionally erroneous, recognition of frames, scripts, and actions. T-rules are less efficient but, when needed, will lead to successful problem-solving. It is difficult to strictly order rules within either set, or across sets, so fuzzy set theory was employed.

The model was used to predict the diagnosis behavior of airframe and power-plant mechanics when troubleshooting simulated automobile engines, turbo-prop engines, jet engines, turbocharged engines, autopilots, and helicopters. The model very closely predicted their overall troubleshooting performance. It predicted their exact choices of actions 50–60 % of the time. The predicted actions were equivalent to humans' actions, in terms of membership in the choosable set, a much higher percentage of the time.

It should be noted that this type of model is inherently very context dependent. This is especially the case for the S-rules. Hence, there is no general off-the-shelf model of human problem-solving. There are generally applicable frameworks, but they must be populated with much context-specific knowledge. Consequently, there is no quick and easy way to computationally model the problem-solving behaviors of humans across domains.

How might these models of human problem-solving have helped with EAL Flight 401? The EAL crew and the crews in our simulator experiment knew that the light should come on when the landing gear are engaged. When the light did not come on, they formed two hypotheses—the gear had not engaged or the bulb had burned out. It was easier to check the light bulb than to fly by the control tower to have the landing gear visually inspected.

Things became much more complicated when they dropped the light bulb and had to search for it amidst the messy landscape of an aircraft cockpit. Finding the bulb captured everybody's attention. One of the crew members bumped the controls, disengaging the autopilot, while searching for the bulb. Finding the bulb preempted all other tasks.

Recognizing the bulb would rely on S-rules. Hypothesizing where the bulb might be would involve T-rules concerning the layout of the cockpit. This thinking captured them. Supervisory control of the automation was ignored. The uncontrolled aircraft spiraled into the swamp. They failed in terms of both problem-solving and control.

Summary

Considering the example modeling capabilities discussed in this section in terms of the representational paradigms discussed in Chapter 2, six are represented:

- dynamic systems theory: the dynamics of the aircraft and ship can be represented by differential or difference equations;
- control theory: the supervisory control tasks for the aircraft and ship can be represented in terms of feedback control, including compensation for failures;
- estimation theory: humans' estimation of the state of the system, as well as detection of failures, can be modeled using filtering theory;
- queuing theory: multitask decision-making for both examples is, in part, a queuing problem, coordinated with the supervisory control problem;
- decision theory: signal detection theory and pattern recognition methods can be used to detect failures; and
- problem-solving theory: rule-based models of problem-solving, for example, S-rules and T-rules, can be used to diagnose failures.

Thus, six of the eight paradigms from Chapter 2 are of use in representing human behaviors in the two examples.

Mental Models

I mentioned the notion of mental models earlier in this chapter. This ambitious construct embraces a broad range of behavioral phenomena. It also has prompted various controversies. On one hand, it would seem that people must have mental models of their cars, for example. Otherwise, how would people be able to so proficiently negotiate traffic, park their cars, and so on?

On the other hand, perhaps people have just stored in their memories a large repertoire of patterns of their car's input–output characteristics, a large look-up table, if you will, of steering wheel angles and vehicle responses. From this perspective, there is no "model" per se—nothing computational that derives steering wheel angle from desired position of the car.

This is a difficult argument to resolve if one needs proof of the representation of the model in one's head. Are their differential equations, neural nets, or rules in the driver's head? Alternatively, one might adopt a functional point of view and simply claim that humans act as if they have certain forms of models in their brains that enable particular classes of behaviors.

We became deeply engaged in this issue, reviewing a large number of previous studies and publishing the highly cited "On looking into the black box: Prospects

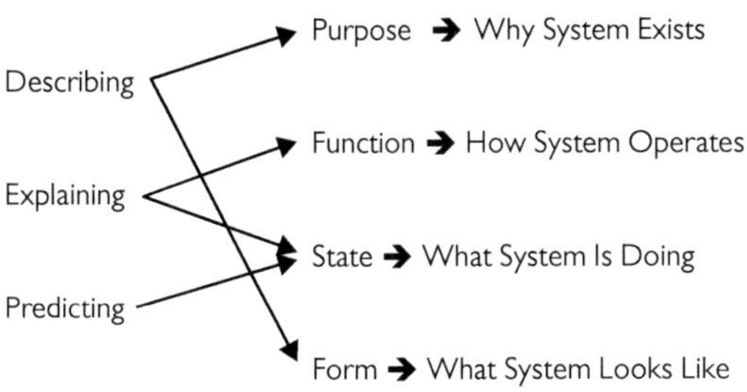

Fig 6.3 Functions of mental models.

and limits in the search for mental models" (Rouse & Morris, 1986). We addressed the basic questions of what do we know and what can we know.

The definition that emerged, summarized by Figure 6.3, was, "Mental models are the mechanisms whereby humans are able to generate descriptions of system purpose and form, explanations of system functioning and observed system states, and predictions of future system states." This definition only defines the function of mental models, not what they look like.

There are many issues associated with identifying mental models. An overriding issue is accessibility. Tasks that require explicit use of the functions depicted in Figure 6.3 are more likely to involve mental models that are accessible, that is, explainable by humans. In contrast, tasks for which these functions are deeply below the surface of behavior are less likely to involve accessible models.

Another issue is the form of representation. Equations, neural networks, and rules are common choices. It is rarely possible to argue, however, that these representations actually reside in peoples' brains. Contemporary brain research might argue for neural networks being the substrate for all alternative representations. However, such an assertion hardly constitutes proof.

Context of representation is another issue. To what extent are mental models general versus context specific? Do you have different models for one car versus another, or just one model with two sets of parameters? People seem, for instance, to know general rules of algebra rather than having a long repertoire of solutions to specific linear combinations of symbols and numbers. We seem much better at learning rules for solving problems rather than memorizing specific solutions. Perhaps, however, we are fooling ourselves in thinking about how we think.

All of the above comes together when we consider the issue of instruction. What mental models should we attempt to create, and how should they be fostered? Do people need to know theories, fundamentals, and principles, or can they just learn practices? The significance of this issue cannot be overstated. Training is discussed later in this chapter.

Team Mental Models

I have thus far addressed individual behavior and performance. Let's now consider groups or teams, which typically perform within a workspace where they can see or, at least, hear each other. As I indicated earlier, one of our studies of teams was prompted by the Aegis cruiser USS *Vincennes* shooting down an Iranian passenger airliner in the Persian Gulf in 1988. We began by observing teams at the Aegis test facility in Moorestown, New Jersey. This training facility is unusual in that it looks like a ship superstructure rising out of a field west of the New Jersey Turnpike. It is sometimes referred to as the "Cornfield Cruiser."

Training exercises involved 25 people who staff the Combat Information Center (CIC). At the time of our observations, there were typically two exercises per day, each of which took considerable preparation, pre-briefing, execution, and debriefing. I was struck by the fact that every exercise seemed to be aptly termed "25 folks face Armageddon." This is what Aegis is designed to do. However, as the July 3, 1988, incident shows, not all situations are Armageddon.

We focused our study on the CIC's anti-air team, as that was the team that dealt with the Iranian airliner. Our initial observation of this team suggested that team members did not have shared mental models. In particular, we hypothesized that inadequate shared models of teamwork—in contrast to mental models of equipment functioning or taskwork—hindered the team's performance (Rouse, Cannon-Bowers, & Salas, 1992).

Our definition of mental models for teamwork followed the functional definition in Figure 6.3. However, the knowledge content, as shown in Table 6.2, differs from that described in our earlier discussions. The overall research questions concerned what elements of Table 6.2 were needed by the anti-air team and how best to impart this knowledge.

The overall conclusions of this research were that the anti-air team members were not well coordinated and did not communicate well in terms of their behaviors in these demanding exercises. It appeared that the members of this team often did not know what was expected of them by other team members and did not know what they could expect of others outside the team. Because of this, they often did not communicate with themselves or with others, or, if

Table 6.2 ***Knowledge content of mental models for teamwork***

Level	Types of Knowledge		
	What	How	Why
Detailed/ Specific/ Concrete	Roles of team members (who member is)	Functioning of team members (how member performs)	Requirements fulfilled (why member is needed)
↓	Relationships among team members (who relates to whom)	Co-functioning of team members (how members perform together)	Objectives supported (why team is needed)
Global/ General/ Abstract	Temporal patterns of team performance (what typically happens)	Overall mechanisms of team performance (how performance is accomplished)	Behavioral principles/ theories (why: psychology, management, etc.)

they did, they did so erroneously or ambiguously. Occasionally, they could not explain or interpret communications received.

It was clear that the anti-air team needed a much-improved shared mental model of teamwork. This led to the development of an intervention called Team Model Training, which is discussed later in this chapter.

Summary

Computing possible futures where behavioral and social phenomena are central presents several challenges. In particular, projecting how people will behave depends on understanding the mental models that determine their behaviors. At one extreme, their mental models will mimic the structure of the task environment to the best of their abilities. At the other extreme, when they have complete discretion, their beliefs, perceptions, and motivations can prompt a wide variety of behaviors.

Application to Failure Situations

It is useful to consider how models of manual control, multitask decision-making, and problem-solving can be conceptually integrated to understand how human operators and maintainers address system failure situations.

Control is the central task when operating a system. This may involve direct manual control or supervisory control of automatic control systems. We can see this spectrum when we look at how automobile driving has evolved from Ford's Model T to Tesla's Autopilot to emerging autonomous vehicles. At some level, likely not in the vehicle, there is a supervisory control function staffed by humans.

Multitask decision-making concerns how humans interleave control tasks and discrete tasks such as communications. Discrete tasks can distract humans from performing control tasks, as evidenced by the fact that texting while driving is linked to an increased accident rate. Nevertheless, coordination of multiple tasks is often inherently necessary, for example, when aircraft pilots need to communicate with air traffic control.

We can think of normal operations as involving this ongoing interleaving of control tasks and discrete tasks. Occasionally, however, problems arise, as the cases discussed earlier in this chapter illustrate. I will term such problems "failures," in the sense that the normal operations fail; there is not necessary a hardware or software failure. There may, for example, be an organizational failure, as exemplified by the Aegis case.

How do operators and maintainers detect failures? First of all, they may not detect the problem, as was the situation for the EAL crew. By the time that they knew something was wrong, it was far too late to recover. The case studies discussed in the next section focus on situations where the personnel involved detected that something was wrong but had difficulty diagnosing the problem and, consequently, were not sure how best to compensate for the problem.

Failure detection can occur in a variety of ways. A blown tire while driving is usually very obvious. On the other hand, it may take a while to notice a "check engine" light. Before smart alarms, it was common in power-plant control rooms for hundreds of alarms to sound simultaneously. The control-room crew's first reaction was often to cancel all the alarms.

Detection difficulties can arise when failures are subtle. One failure we studied was in a supertanker power-plant simulator. The crew would notice that the boiler level was full but the feedwater valve was wide open. Why would water be flowing into the boiler when it was already full? They knew something was wrong, but the symptoms were confusing.

This led us to identify a deficiency in the crew's training. They had been taught about how the control systems worked, but not about how they could fail. In this case, the boiler level sensor had failed, thereby providing a signal to the control system that the boiler was empty. Once the implications of sensor failures were added to the course content, crews were readily able to detect and diagnose what happened.

Diagnosis involves figuring out the source of the failure symptoms detected. Pressing the start button in the car results in a click but no start. The battery is dead. The battery needs to be charged, which you do. The next day, you have the same problem. What's draining the battery? Much tracing and testing identifies a short in the lighting circuit. Taping the frayed wire solves the problem.

Sometimes, S-rules (symptomatic rules) enable diagnosis. Noise patterns, for example, clearly identify the culprit. When S-rules do not work, you need T-rules (topographic rules) that address the structure of the system. These rules enable tracing the structure and determining the most diagnostic tests. This is elaborated in the next section.

Once the failure has been detected and diagnosed, the next task is compensation. Continued system operation may require that the failure be fixed, but there are times when operation can be continued despite the failure. For example, many airplanes can be flown acceptably despite the loss of an engine. They have to be trimmed somewhat differently, but pilots are trained to know how to do this.

Case Studies

The case studies in this section focused on failure diagnosis. The types of models discussed thus far in this chapter can be used to predict human behavior and performance when diagnosing failures. In this section, I discuss how these models can play central roles in augmenting human behavior and performance. I consider two ways to augment humans—aiding and training.

Aiding

Aiding is concerned with directly augmenting human behavior and performance. We can use models such as those depicted in earlier figures to understand human behavior and performance—in real time as it is happening. This understanding can be used by the aiding subsystem to assist humans by tailoring their displays and controls, monitoring behaviors, and adapting assistance according to humans' needs and desires. I will defer elaboration of these ideas until Chapter 8 on intelligent systems (Rouse, 2007, 2015; Rouse & Spohrer, 2018).

Training

Training is concerned with creating the potential to perform by enabling people to acquire the knowledge and skills needed for the tasks of concern. As the focus of this chapter is system failures, one objective is to train personnel to detect, diagnosis, and compensate for failures, in both maintenance and operations. A second objective is to train personnel to coordinate performance in team-based environments.

There are several things needed to develop training for such tasks. First, we need a model of the system of interest, for example, the automobile in Figure 6.4. This example is much less detailed than what we actually used, which needed enough detail to be able to identify failed components. Nevertheless, this figure exemplifies the idea. Associated with this figure is a more abstract model that enables predicting the consequences of each block in Figure 6.4 failing, with variations for hard failures versus degradations. This model also has to project how these consequences propagate to create symptoms.

Beyond these models of the system, we need models of the tasks, that is, the workflow associated with operating and maintaining the system. For team-based environments, this includes, at least, the upper left elements of Table 6.2. Thus, we need both task and team models.

Another set of models represents diagnostic strategies and tactics. There are many context-specific nuances, but two strategies broadly apply:

- When forming hypotheses, first consider what has not failed; for example, I have lost Internet connectivity, but not cable television—therefore, the problem must be in the router, not the cable box. This strategy tends to significantly reduce the size of the feasible set of failures.

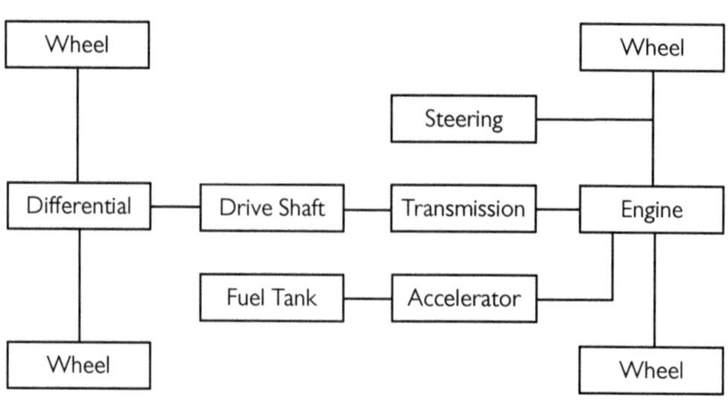

Fig 6.4 Automobile model.

- When testing members of the feasible set, choose tests that will determine which half of the feasible set contains the failure, perhaps weighted by the a priori probabilities of each element of the set being the failure. This is referred to as "half-split testing."

We devised an approach to computer-based training that we termed Simulation-Oriented Computer-Based Instruction (SOCBI), which simulated systems such as the one depicted in Figure 6.4 and provided instruction to help trainees understand and internalize strategies such as those listed above. We deployed and evaluated SOCBI at the Institute of Aviation at the University of Illinois for airframe and power-plant mechanics, at Marine Safety International for supertanker engineers, and at the U.S. Army's Fort Gordon, for the Signal Corps.

Studies of these training systems at these venues and several others provided a variety of insights. First, it is difficult to teach abstract strategies such as elaborated above. Trainees say they understand the strategies but then often do not employ them. We found two ways to circumvent this.

One was to provide aiding during the training. Specifically, the computer would automatically apply the first strategy and display what remained in the feasible set as testing progressed. This greatly improved trainees' performance in terms of reduced time and errors, for example, making useless tests. These improvements remained after the aiding was removed.

Second, we created problem scenarios where trainees would discover the principles, such as the second one above. For scenarios where the feasible set was a single stream of connected components, half-split testing was a fairly intuitive approach. As this was mastered, the connections were incrementally made more complicated, but trainees would continue with half-splitting. It was common for them to report to us their discovery of this principle.

In general, we found that novice trainees had more difficulty absorbing the more abstract principles. They were still trying to learn the basics of the training context. Advanced trainees, for example, those in their second year, benefited more substantially from the training techniques just described.

Mixed-Fidelity Training

Our experiences with SOCBI made it clear that immersing novice trainees in full-scope training simulations was often overwhelming. The concept of mixed-fidelity training, shown in Figure 6.5, emerged. Put simply, the low-fidelity

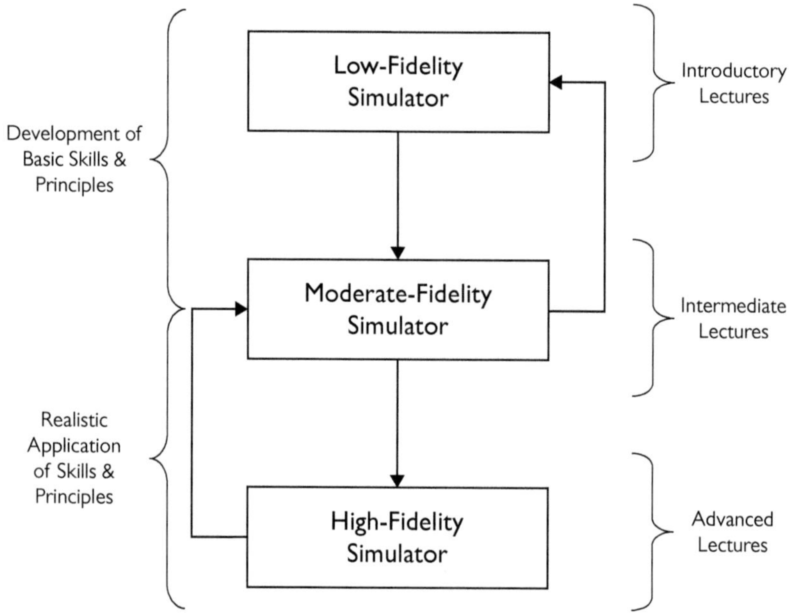

Fig 6.5 Mixed-fidelity training.

simulator, and associated instruction, prepared trainees to perform in the moderate-fidelity simulator, which, in turn, prepared them to perform in the high-fidelity simulator.

Beyond the pedagogical value of this approach, it also better managed the flow of trainees. The low- and moderate-fidelity simulators were typically hosted on personal computers. Thus, for example, one could have a dozen trainees using these simulators while one or two were using the high-fidelity simulator. This was the case for Marine Safety International as well as for Aegis training.

The Aegis training involved half-day large-scale exercises with 25 people in the simulated CIC. Our observations were that team members did not understand what each other was doing and what each team member expected of each other team member. Consequently, the training exercises were often rather chaotic.

Team Model Training was a desktop computer-based training system that approximated the full-scope Aegis simulator. In it, the user would participate as a member of a team where all other team members were simulated. The user would both learn their own role and, by playing other team members' roles, learn how to fit into the team's performance. Users could easily complete 20–30 exercises before venturing into the full-scope simulator.

An evaluation of Team Model Training compared it with more conventional lecture-based training. Performance was assessed in the full-scope simulator after the training methods had been completed. Team Model Training significantly decreased the overall number of communications made by team members, indicating that well-trained teams are not necessarily those with the highest numbers of explicit communications. Specifically, frequent requests for clarifications of communications, while addressing missiles incoming in the next few seconds, should be avoided.

We also addressed team coordination and training in the performing arts (Rouse, 2007). Space does not allow relating this study. We found, however, that coordination and training are strongly influenced by the role of the leader and the familiarity of team members, which varied across the 12 performing arts studied. For example, if the leader prepares the team but does not participate in the performance, coordination is much more explicitly programmed.

Conclusions

System failures present an interesting scenario for predicting possible futures. Human detection, diagnosis, and compensation for failures are often expected of personnel associated with operating and maintaining the systems of interest. Increasingly, this is the primary purpose of these personnel. Automation is typically employed for normal and routine operations.

Concerns include the extent to which humans will detect that the nonroutine has happened, correctly diagnose what has gone wrong, and devise appropriate means for compensating for the system's degraded state. We understand humans' abilities, limitations, and inclinations in such situations. With well-designed training and aiding, humans can fill this role quite well. Neither the EAL L-1011 crew nor the Aegis cruiser crew had this training and aiding.

This chapter has been rather different from earlier chapters. The question of interest involved preparing personnel to perform in future failure situations. We predicted their likely behaviors and performance and then evaluated the extent to which these predictions were reasonable, in terms of both diagnostic performance and team coordination. The result was the development of several training innovations that were subsequently deployed.

Key Points

- Addressing system failures is an important aspect of operating and maintaining complex systems, particularly when the systems are laced with behavioral and social phenomena.
- Despite advances in technology, humans will inevitably have roles in addressing failures when detection, diagnosis, and compensation cannot be automated.
- Human problem-solving involves a mix of pattern recognition and structural sleuthing based on mental models for taskwork and teamwork.
- Training and aiding can enhance human problem-solving performance by fostering problem-solving strategies and tactics, as well as team coordination.

REFERENCES

Johannsen, G., & Rouse, W. B. (1983). Studies of planning behavior of aircraft pilots in normal, abnormal, and emergency situations. *IEEE Transactions on Systems, Man and Cybernetics, SMC-13*(3), 267–278.

Klein, G. (2004). *The power of intuition: How to use your gut feelings to make better decisions at work*. New York: Currency.

Minsky, M. (1975). A framework for representing knowledge. In P. H. Winston (Ed.), *The psychology of computer vision* (pp. 211–277). New York: McGraw-Hill.

Newell, A., & Simon, H. A. (1972). *Human problem solving*. Englewood Cliffs, NJ: Prentice-Hall.

Pew, R. W., & Mavor, A. S. (Eds.). (1998). *Modeling human and organizational behavior: Applications to military simulations*. Washington, DC: National Academy Press.

Rasmussen, J. (1983). Skills, rules, and knowledge; signals, signs, and symbols, and other distinctions in human performance models. *IEEE Transactions on Systems, Man and Cybernetics, SMC-13*(3), 257–266.

Rasmussen, J. (1986). *Information processing and human–machine interaction*. New York: North-Holland.

Rasmussen, J., Pejtersen, A. M., & Goodstein, L. P. (1994). *Cognitive systems engineering*. New York: Wiley.

Rasmussen, J., & Rouse, W. B. (Eds.). (1981). *Human detection and diagnosis of system failures*. New York: Plenum Press.

Rogers, S., Rogers, W., & Gregston, G. (1992). *Storm center: The USS Vincennes and Iran Air Flight 655: A personal account of tragedy and terrorism*. Annapolis, MD: Naval Institute Press.

Rouse, W. B. (1980). *Systems engineering models of human–machine interaction.* New York: North Holland.
Rouse, W. B. (1983). Models of human problem solving: Detection, diagnosis, and compensation for system failures. *Automatica*, 19(6), 613–625.
Rouse, W. B. (2015). *Modeling and visualization of complex systems and enterprises: Explorations of physical, human, economic, and social phenomena.* New York: Wiley.
Rouse, W. B. (2007). *People and organizations: Explorations of human-centered design.* New York: Wiley.
Rouse, W. B., Cannon-Bowers, J. A., & Salas, E. (1992). The role of mental models in team performance in complex systems. *IEEE Transactions on Systems, Man, and Cybernetics, 22*(6), 1296–1307.
Rouse, W. B., & Morris, N.M. (1986). On looking into the black box: Prospects and limits in the search for mental models. *Psychological Bulletin, 100*(3), 349–363.
Rouse, W. B., & Rouse, R. K. (2004). Teamwork in the performing arts. *Proceedings of the IEEE, 92*(4), 606–615.
Rouse, W. B., & Spohrer, J.C. (2018). Automating versus augmenting intelligence. Journal of Enterprise Transformation. doi:10.1080/19488289.2018.1424059
Schank, R. C., & Abelson, R. P. (1977). *Scripts, plans, goals and understanding.* Hillsdale, NJ: Lawrence Erlbaum.
Sheridan, T. B. (1992). *Telerobotics, automation and human supervisory control.* Cambridge, MA: MIT Press.
Sheridan, T. B. (2017). *Modeling human–system interaction: Philosophical and methodological considerations, with examples.* Hoboken, NJ: Wiley.
Sheridan, T. B., & Ferrell, W. R. (1974). *Man–machine systems: information, control, and decision models of human performance.* Cambridge, MA: MIT Press.
Zacharias, G. L., MacMillan, J., & Van Hemel, S. B. (Eds.). (2008). *Behavioral modeling and simulation.* Washington, DC: National Academy Press.

CHAPTER 7

Health and Well-Being

In the last chapter, I addressed the health and well-being of complex engineered systems. This chapter focuses on the health and well-being of people. I consider how health services are provided in the U.S. delivery ecosystem. This highly fragmented system presents challenges with providers, payers, and regulators at local, state, and federal levels. Not surprisingly, the emphasis of this chapter is on model-based approaches to support decision-making.

Two questions have driven the six case studies discussed. First, how can innovative pilot studies, involving perhaps 500–1,000 patients, be scaled for delivery to hundreds of thousands of patients? The focus of the case studies discussed here included diabetes, heart disease, Alzheimer's disease, and transition care for complex elderly patients.

The second question concerned broad enterprise-level issues. How is the health ecosystem of New York City likely to evolve in the next decade or two? How can the overall health of a broad population be maintained? How is technology likely to drive the quality and costs of health? Addressing such questions requires ambitious modeling efforts.

These scaling and enterprise issues are far from new. They have long been addressed via empirical studies, advisory boards, and other traditional mechanisms. This chapter addresses these issues computationally. Possible futures are computed for alternative current investment decisions. We compute what might happen and the conditions under which possible futures are likely.

Computing Possible Futures. William B. Rouse. Oxford University Press (2019). © William B. Rouse 2019.
DOI: 10.1093/oso/9780198846420.001.0001

Delivery Ecosystem

Figure 7.1 portrays who is involved in providing health, education, and social services in the United States, and the inherent difficulty of accessing these services. The ecosystem is highly fragmented, often resulting in low-quality, expensive services. Equity of services is also a major issue, with social determinants of health having enormous impacts.

Our approach employs a framework we have developed for modeling complex social enterprises (Rouse, 2015) and applied in domains ranging from healthcare delivery (Rouse & Cortese, 2010; Rouse & Serban, 2014) to higher education (Rouse, 2016). This framework addresses the physical, human, economic, and social phenomena underlying complex ecosystems. A population health version of this framework is shown in Figure 7.2.

This multilevel framework provides the basis for integrating different types of computational models to explore policy alternatives. The people level is usually agent based, laced with both decision theory and behavioral economics. The processes level is represented as networks of flows, and the resulting queues. The organizations level involves the microeconomics of resource allocation, again laced with both decision theory and behavioral economics.

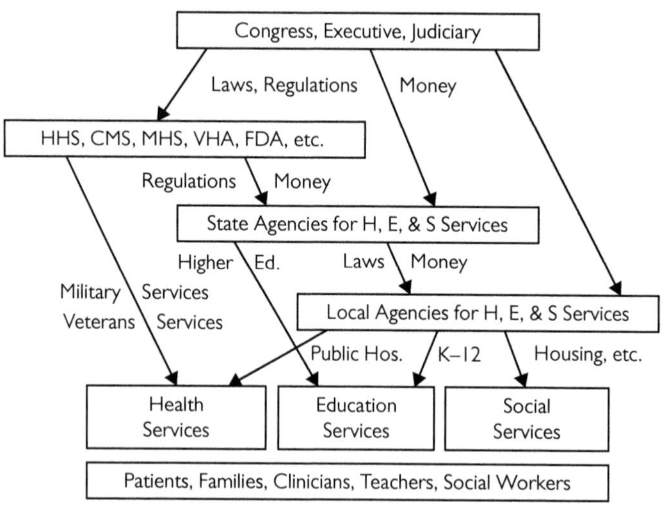

Fig 7.1 Relationships among organizations and services; CMS: Centers for Medicare and Medicaid Services; FDA: Food and Drug Administration; H, E, & S: health, education, and social services; HHS: U.S. Department of Health and Human Services; Hos.: Hospitals; MHS: Military Health System; VHA: Veterans Health Administration.

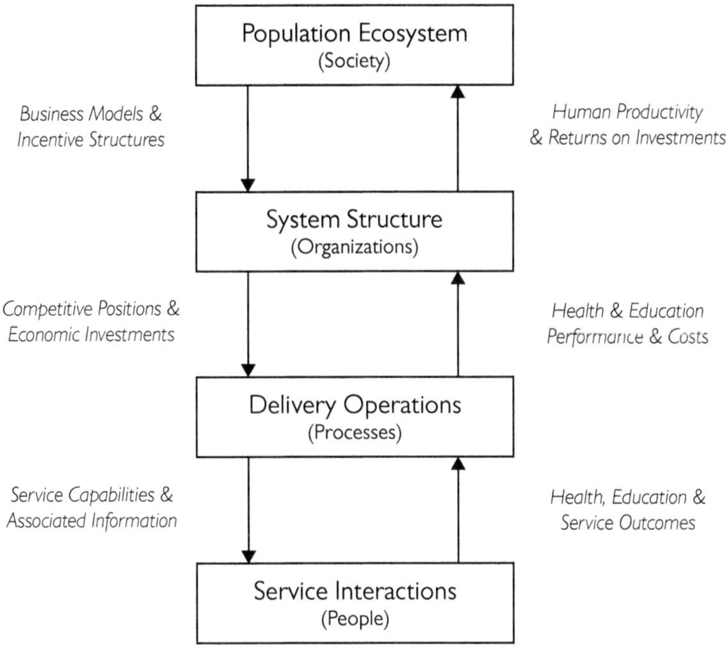

Fig 7.2 Population health enterprise.

The society level involves the macroeconomics of policy. The resulting multi-level model is typically embedded in an interactive visualization that enables experimentation.

At the people level, central phenomena include establishing a route through the many needed services. People may balk (not become patients) or renege (drop out of treatment) along the route, due to delay times and other factors.

Processes-level phenomena include getting appointments for each service in the route. Delays between services can be characterized in weeks. Delays are highly affected by capacity constraints. Processes also involve the flow of information among service providers. Inefficiencies in the flows of information can foster inefficiencies in the flow of patients to services.

At the organizations level, capacity constraints are due to investment policies, as well as availability of personnel. Organizations, not surprisingly, tend to invest in capacities needed to provide services that are highly reimbursed. Thus, for example, cancer, cardio-, and ortho-surgery services are typically better provisioned than chronic disease management is.

On the society level, investment policies are related to payer reimbursement policies for different services. This level also relates to how value is defined. Healthy people not only have lower healthcare costs; they also typically work, earn incomes, pay taxes, and so on. Thus, society benefits from a healthy population far in excess of the lower healthcare costs. Fragmentation at the highest level undermines accounting for the full benefits of population health.

Figure 7.2 provides a framework for considering the investments and innovations needed to enable integrated care, as well as the implications of such initiatives. The following six case studies build upon this framework.

Scaling Innovation

The three case studies in this section concern computational models for studying the issues associated with scaling successful pilot studies or randomized clinical trials to broad adoption. In general, inefficiencies not visible with a few hundred patients can become glaring for hundreds of thousands of patients.

Emory Prevention and Wellness

This case study addressed the employee prevention and wellness program of Emory University (Park, et al., 2012). The application of the multilevel model focused on the roughly 700 people in this cohort and their risks of diabetes mellitus (DM) and coronary heart disease (CHD). Each person's risk of each disease was calculated using DM and CHD risk models from the medical literature, using initial individual assessments of blood pressure, fasting glucose level, and so on. Subsequent assessment data were used to estimate annual risk changes as a function of initial risks of each disease.

The model of this healthcare delivery enterprise included the four levels shown in Figure 7.2—society (called the "ecosystem level" in this model), organizations (called the "organization level" in this model), processes, and people. Each level introduces a corresponding conceptual set of issues and decisions for both the payer and the provider. In this case, the Human Resources Department of Emory University (HR) was the payer responsible for healthcare costs for university employees, while the Predictive Health Institute (PHI) was the provider focused on prevention and maintenance of employee health.

The ecosystem level allows decision-makers to test different combinations of policies from the perspective of HR. For instance, this level determines the allocation of payments to PHI based on a hybrid capitated and pay-for-outcome formula. It also involves choices of parameters such as projected healthcare inflation rate, general economy inflation rate, and discount rate that affect the economic valuation of the prevention and wellness program. One of the greatest concerns of HR is achieving a satisfactory return on investment (ROI) on any investments in prevention and wellness.

The concerns at the organization level include the economic sustainability of PHI—their revenue must be equal to or greater than their costs. To achieve sustainability, PHI must appropriately design its operational processes and rules. Two issues are central. What risk levels should be used to stratify the participant population? What assessment and coaching processes should be employed for each strata of the population? Other organization-level considerations include the growth rate of the participant population, the age ranges targeted for growth, and the program duration before participants are moved to "maintenance."

The process level represents the daily operations of PHI. Participants visit PHI every 6–12 months. Seven health partners employed by PHI perform assessments, work with participants to set health goals, and perform follow-up calls or emails to monitor participants and encourage them to follow their plan. All of these activities are captured in the processes level. The costs of these activities are aggregated and reflected in the organization level as the costs of running PHI.

The people level is the replication of the actual population of PHI participants. Over a three-year period, roughly 700 participants joined this prevention and wellness program. Each of them had various assessment measurements recorded, such as blood pressure, fasting glucose level, and so on; because PHI is, in part, a research project, approximately two thousand variables were measured at each assessment encounter. Each participant was instantiated as an agent in the model. Based on the assessment measurements, the risk of developing DM or CHD was computed for each agent. Then, total healthcare costs were estimated for the participants' remaining life, based on their risk level for each disease. The reduced amount of aggregated total healthcare cost achieved by PHI is an ecosystem-level benefit for the HR organization.

Runs of the multilevel simulation are set up using the dashboard shown in the screenshot in Figure 7.3. Beyond the decision variables discussed above, decision-makers can decide what data source to employ to parameterize the models—either data from the American Diabetes Association (abbreviated as ADA in the figure) and the American Heart Association (abbreviated as AHA

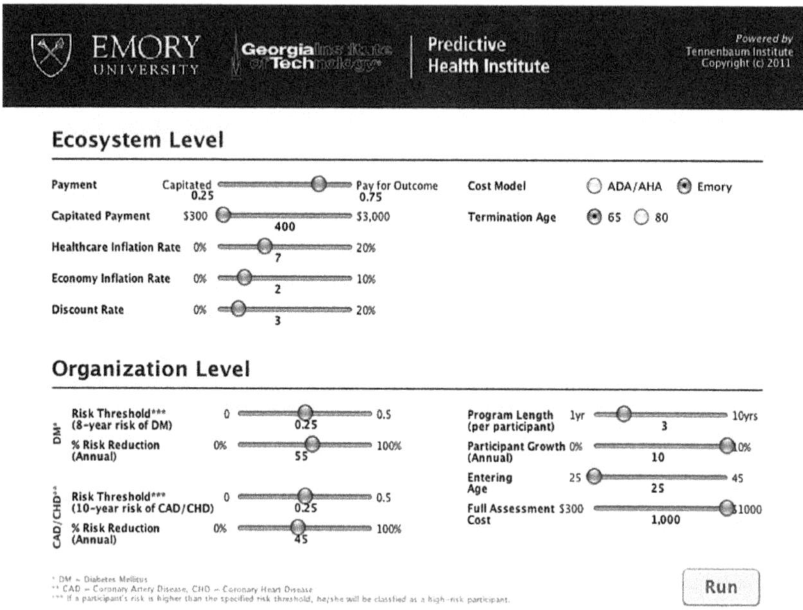

Fig 7.3 Multilevel simulation dashboard; ADA: American Diabetes Association; AHA: American Heart Association.

in the figure), or data specific to Emory employees. Decision-makers can choose to only count savings until age 65 or also project postretirement savings.

The bottom half of the dashboard provides inputs from organization-level decision-makers, namely PHI. Beyond the variables mentioned above, these decision-makers must choose how to stratify the participant population into low- and high-risk groups for each disease. Once they choose a level on the "Risk Threshold" slider, a set point appears on the "% Risk Reduction" slider that represents what PHI is actually achieving based on analysis of their ongoing assessment data. Decision-makers can choose to operate at the set point by moving the slider to this point, or they can explore the consequences of larger or smaller risk reductions.

Figure 7.4 contains a screenshot showing the ecosystem and organization levels of the model. The provider organization, PHI, decides how to stratify participant flows and seeks to have revenues equal or exceed costs. The payer organization, HR, sets the "rules of the game" as depicted on the dashboard in Figure 7.3. HR's ROI from PHI's services is shown in NPVs, using the discount rate shown in Figure 7.3. The returns achievable with various combinations of the parameters were explored during numerous meetings with Emory.

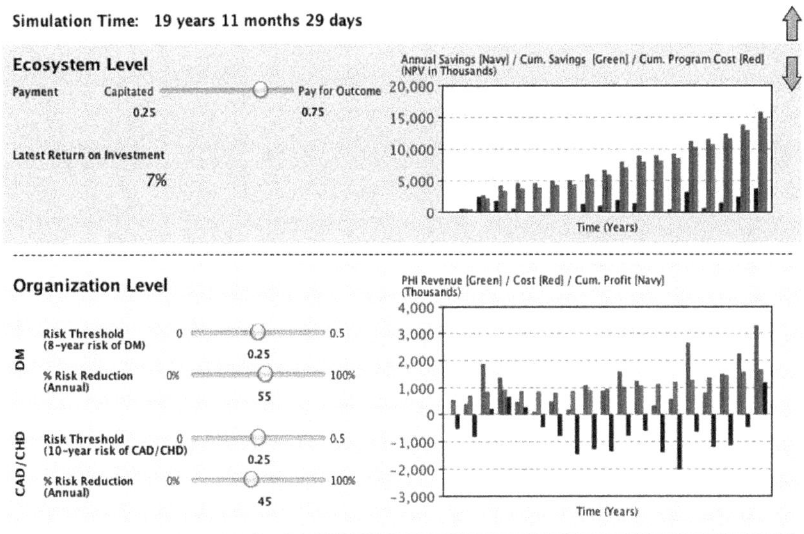

Fig 7.4 Ecosystem and organization levels of models; CHD, coronary heart disease; Cum.: cumulative; DM: diabetes mellitus; NPV: net present value; PHI: Predictive Health Institute.

Evaluating PHI as it was currently operating resulted in several conclusions. If Emory scales up its current prevention and wellness program, it will yield an annual −96 % ROI—about as bad as you can get. Yet, they are achieving dramatic improvements for people with high risks of diabetes and heart disease. If we radically reorganize this program (computationally), we can achieve a 7 % ROI for Emory and sustain the program. Those are the results through age 65.

If we extend the analysis to age 80, we achieve a 30 % ROI. This difference is all to the benefit of the U.S. Centers for Medicare and Medicaid Services (CMS). Thus, in principle at least, CMS should incentivize Emory to the provide prevention and wellness to its employees. In general, CMS should incentivize all employers to keep people healthy so that when they enter Medicare at age 65, they are much healthier and less expensive for CMS.

To achieve this impressive ROI and stay in business, PHI will have to change its business model, stratifying the population by risk levels and tailoring processes to each stratum. This could include an initial low-cost, streamlined assessment and subsequently PHI "Lite" for low-risk participants. PHI also needs to develop a low-cost "maintenance" process to sustain reduced risks once they have been achieved. These recommendations significantly influenced the subsequent redesign of PHI.

Indiana Alzheimer's Care

This case study addressed care for patients with memory and emotional problems such as Alzheimer's disease and other related types of dementia. There is a substantial need to develop new scalable and sustainable brain-care services to care for these patients. This care requires extensive psychosocial support, nursing care, and comprehensive patient-centered management, which strain the resources of clinicians, family caregivers, and community-based support structures. Indiana University developed such a health management program, which was called the Aging Brain Care Medical Home (ABC), to provide the collaborative care model to 1,500 older adults in central Indiana (LaMantia et al., 2014).

In order to scale up the ABC collaborative care model to more patients and other geographical areas, it was necessary to understand what factors affect the efficiency and effectiveness of its operations and outcomes. To this end, we developed a multilevel computer simulation model of the ABC program (Boustany, Pennock, Bell, Boustani, & Rouse, 2016). It was shown that scaling the program without modification would be infeasible. More broadly, the ABC simulation model served as a risk reduction decision support tool for healthcare delivery redesign, allowing early identification of operational and redesign issues.

The ABC simulation model included elements from both agent-based and discrete-event modeling, which were incorporated in the overall multilevel model. The model was used to explore different strategies for scaling up the ABC program. Results showed that as population sizes increase, economies of scale are reached, and thus the contribution of fixed costs to the costs per member or per patient decreases. Another important finding that emerged from this study was that the impact of the ABC program on cost savings reaches a steady state after a period of several years, which is indicated by a decreasing standard error and plateaued ROIs.

In the process of conducting this study, we encountered a data set for 70,000 Alzheimer's patients over several years. This allowed is to estimate transition rates among six states of health: normal, mild cognitive impairment, mild Alzheimer's, moderate Alzheimer's, severe Alzheimer's, and death. We also had data on the annual costs of care for each state.

Alzheimer's disease cannot be cured or reversed at this time. However, progression can be delayed via various interventions. We explored the impacts of delays, by varying probabilities of retaining patients in less advanced states of

0 %, 5 %, or 10 %. Using simulation, we found that a 5–10 % increase in retaining patients in less advanced states can yield enormous annual savings of roughly 50 % by Year 6. The percent savings increases with age, because older patients die before they advance to the severe stage and need nursing home care.

This use of modeling is interesting in that we are predicting the economic value of an unknown intervention. We do not know how to increase the probabilities by 5–10 %. However, we do know what it would be worth if we could develop this intervention—half the cost of Alzheimer's care in the United States. Our goal was to motivate parties who might be able to figure this out.

Penn Transition Care for Elderly Patients

The Transitional Care Model (TCM) is a proven care management approach that can contribute to a more person-centered, effective, and efficient response to the challenge of chronic illness, including needs to avoid readmissions and associated penalties. Despite TCM's proven value (Naylor, 2012; Naylor et al., 2014), it has been challenging to convince decision-makers to implement this model. Success in TCM's spread has been achieved only slowly—one health system or community at a time. Among major barriers to widespread implementation are the perceptions that the model has been demonstrated to work in randomized control and comparative effectiveness trials but not in the "real world"; that it is too complex and costly, requiring upfront investment which will largely benefit other providers downstream; and that it is not adaptable to local contextual issues.

While each of these misperceptions have been addressed through successful translation of the TCM in a number of health systems, traditional strategies (e.g., identifying local champions, multiple meetings with decision-makers) consume substantial amounts of time and are not as efficient as desired in promoting widespread scaling. Such challenges are not limited to the adoption of the TCM, and addressing them could have positive impacts on the widespread adoption of evidence-based care throughout the U.S. healthcare system.

To that end, the specific goal of this case study was to determine whether the use of a policy flight simulator accelerates positive decisions to implement the TCM (Pennock et al., 2018; Rouse, Naylor, et al., in press). As indicated in earlier discussions throughout this book, policy flight simulators fuse aspects of scientific analysis, engineering, social science, and visualization to provide decision-makers with a more comprehensive understanding of the consequences of interventions than that provided by traditional mathematical and computational approaches.

To accomplish this goal, the team conducted two activities in an iterative, adaptive process. First, we elicited barriers and facilitators to adopting evidence-based, highly effective interventions from decision-makers representing providers, payers, and purchasers. Second, we developed and continuously refined the TCM policy flight simulator, building upon Figure 7.2 and creating interfaces similar to those shown in Figures 7.3 and 7.4. In the process, several key insights emerged:

- The payment system is central.
- Beliefs about evidence vary; peers' actions are important.
- Research evidence is not sufficient.
- The offering must relate to "my population."

These insights caused us to realize that any investment decision of the magnitude of TCM would likely require the involvement of many stakeholders and organizations in a given healthcare system. Consequently, we elaborated our goal, namely, to determine whether the use of an innovative policy flight simulator would help healthcare decision-makers (providers, payers, or purchasers) make better-informed decisions regarding the adoption of TCM and increase their confidence in a decision to adopt TCM.

Results showed that we have demonstrated the potential value for a policy flight simulator to inform decisions about adopting evidence-based interventions. The TCM simulator enables providers and payers to project the impact of TCM on their patient populations, using their financial parameters, for example, local wages. This increases their confidence in how this evidence-based intervention will likely impact them and decreases the tendency to dismiss evidence that the simulator shows to be well-founded.

There is little, if any, doubt that TCM benefits patients across a wide range of patient demographics. Indeed, analysis of benefits by patient demographic characteristics, as reported in studies cited earlier, fails to show any variation in effectiveness with these characteristics. The question addressed here was the extent to which TCM would be economically attractive for any and all providers. The answer is that one size does not fit all. Benefits depend on the patient population enrolled in TCM, the extent of readmission penalties, and the nature of the provider, for example, tertiary versus secondary care and the payment model (capitated vs. fee for service).

As the Medicare population grows and new payment models are deployed, providers will have to understand in depth how their practices affect the economics of their enterprise (Goldsmith & Bajner, 2017), and policymakers will want to know how to anticipate those responses. Indeed, CMS will need to understand that not all providers are the same, that policies that work for one subset of the population may not work for other subsets, and which subsets of the providers or populations are most impacted. Policy flight simulators can provide such understanding before policies are deployed.

Summary

Considering the capabilities discussed in this section in terms of the representational paradigms discussed in Chapter 2, five are represented:

- dynamic systems theory: models of disease incidence and progression, and models of economic growth and inflation;
- queuing theory: patient and clinician use of care process capacities;
- network theory: flow and optimization of services;
- decision theory: microeconomics of investment decision-making, and clinician and patient decision-making; and
- finance theory: discounted cash flows.

Thus, five of the eight paradigms from Chapter 2 were of use in representing the elements and operations of care processes.

Enterprise Models

The enterprise issues discussed in this section are not concerned with particular diseases or types of patients. Instead, the focus is on the broad delivery enterprise, or ecosystem of enterprises, and its behavior and performance. Hospital consolidation, population health, and the impacts of technological innovation are addressed.

New York City Health Ecosystem

The Affordable Care Act is causing a transformation of the healthcare industry. This industry involves complicated relationships among patients, physicians, hospitals, health plans, pharmaceutical companies, healthcare equipment companies, and government. Hospitals are uncertain about how they should best respond to threats and opportunities. This is particularly relevant for hospitals located in competitive metropolitan areas such as New York City, where more than 50 hospitals are competing—many among the nation's best. Questions that arise in this uncertain environment include the following:

- What if we wait until the healthcare market stabilizes and only invest in operational efficiency?
- Should we merge with competing hospitals to increase negotiation power?
- Shall we only focus on acquiring physician practices in highly reimbursed diagnostic groups?

In this case study, we developed a data-rich agent-based simulation model (Yu, Rouse, Serban, & Veral, 2016) to study dynamic interactions among healthcare systems in the context of merger and acquisition (M & A) decision-making. By "data-rich" model, we mean one using extensive rule sets and information sources, compared to traditional agent-based models. The computational model includes agents' revenues and profitability (i.e., financial statements), operational performance, and resource utilization, as well as a more detailed set of objectives and decision-making rules to address a variety of what-if scenarios.

We applied our modeling approach to the M & A dynamics of hospitals in New York City, informed by in-depth data on 66 hospitals in the hospital referral regions for the Bronx, Manhattan, and Eastern Long Island. The objective of the simulation model is to assist hospital executives to assess the impact of implementing strategic acquisition decisions at the system level. This is accomplished by simulating strategies and interactions based on real historical hospital balance sheets and operational performance data.

The outcomes of the simulation include the number of hospitals remaining in the market, and frequent M & A pairs of hospitals under various settings. By varying strategy inputs and relevant parameters, the simulation can be used to generate insights as to how these outcomes would change under different scenarios. The interactive visualizations complement the simulation model by allowing nontechnical users to interactively explore relevant information and input parameter values for different scenarios, as well as view and validate the results of the simulation model.

The results from the simulation model facilitate M & A decision-making, particularly in identifying desirable acquisition targets, aggressive and capable acquirers, and frequent acquirer–target pairs. The frequencies of prevalent pairs of acquirer and target appearing under different strategies in our simulation are of particular interest. The frequency level is a relative value in that it depends on number of strategies included and hospitals involved. A high frequency suggests a better fit and also repeated attraction.

Validation of agent-based simulations is challenging, especially for high-level strategic decision simulations. The overall model and the set of visualizations were validated in two ways. First, from a technical perspective, we compared our simulation results with Capital IQ's data set of hospital mergers and acquisitions transactions. Although there are a limited number of cases under our regional constraint in Capital IQ's database, the realized M & A transactions appear in our results.

Second, we examined the feedback from users. We conducted many (roughly 30) demonstrations for hospital decision-makers and healthcare consultants as well as senior executives from insurance, government, foundations, and so on. In total, perhaps 200 people attended the demos, and many took the controls themselves and tried various options. They made many suggestions, and the number of types of interactive visualizations iteratively increased.

Two predictions resulting from this model were of particular interest: we correctly predicted the Mt. Sinai acquisition of Beth Israel, but we incorrectly predicted that Mt. Sinai would acquire Staten Island University Hospital (Northwell acquired Staten Island instead). During a demo to Mt. Sinai, when we mentioned the latter prediction, a Mt. Sinai executive responded by telling us that Staten Island had been at the top of their list for acquisition but that Northwell had acted more quickly. So, it turns out that this prediction had not been that far off after all.

The key value of the overall model and its set of visualizations is, of course, provided by the insights gained by the human users of this environment—see Figure 7.5. For example, they may determine the conditions under which certain outcomes are likely. They can then monitor developments to see if such conditions are emerging. Thus, they know what *might* happen, even though they cannot be assured of what will happen. The greatest insights are gained not only from simulation, but also from interactive visualizations that enable massive data exploration, which moves from a "one-size-fits-all" static report to more adaptable and useful decision process.

Population Health

Population health involves the integration of health, education, and social services to keep a defined population healthy, address health challenges

Fig 7.5 Simulation of the New York City health ecosystem.

holistically, and assist with the realities of being mortal (Rouse, Johns, & Pepe, in press). The fragmentation of the U.S. population health delivery system makes this very difficult. We are pursuing several questions related to population health and learning health systems (Rouse, Johns, et al., 2017, in press):

- To what extent can upstream interventions (for example, education and social services) decrease the incidence and progression of disease so that the downstream savings justify the upstream investments?
- How sensitive are results to the extent of population engagement? How is engagement affected by payment mechanisms? How might social influences and social media enhance engagement?
- How valuable are the second-order impacts of population health in terms of longer, more productive lives? Are there higher-order impacts that could have significant positive or negative effects?
- To what extent does the fragmentation of delivery, payment, and regulation across local, state, and federal stakeholders undermine the effectiveness of population health offerings? If such fragmentation was not a problem, how well could population health offerings perform in terms of health outcomes and costs?

We addressed these questions in the context of the population health enterprise in Figure 7.2 (Rouse, Johns, et al., s2019). At each level, we considered innovations needed that could leverage systems science, behavioral economics, and social networking. We also considered the implications for health IT and governance. The result was an agenda for transforming population health. An overall computational model that includes the elements of this agenda is very much a work in progress.

Cognitive assistants are likely to play a major role in overcoming fragmentation, in the sense of patients and clinicians feeling that they are using an integrated system despite the remaining organizational silos. I return to this topic in Chapter 8.

Technology Innovation

The last four decades have seen enormous increases in healthcare costs. Specifically, *real* healthcare costs have tripled as a percent of the GDP in the period 1965–2005, with half of this growth due to technological innovation (CBO, 2008). There seems to be virtually unanimous agreement that something has to change significantly.

Figure 7.6 summarizes the overall phenomena discussed in the CBO report. Technological inventions become market innovations as they increase in effectiveness, and associated risks decrease. This results in increased use, which leads to increased expenditures. In parallel, increased efficiency via production learning leads to decreased cost per use, although not enough to keep up with growing use rates in healthcare. Finally, increased use yields improved care that leads to longer lives and increased chances of again employing the technology of interest. For example, the average number of hip replacements in the United States, for people who have hip replacements, is greater than 2.

The concern is how to control the phenomena depicted in Figure 7.6. More specifically, what efficiencies must be realized to steadily decrease the cost per use to offset the constantly increased number of uses, and thereby enable affordable healthcare? We approached this control problem with a series of models, beginning with a very simple model and elaborating as the limits of each model become clear (Rouse & Serban, 2014). The overall conclusion is that costs per use must dramatically decrease to overcome the exponential growth of number of uses.

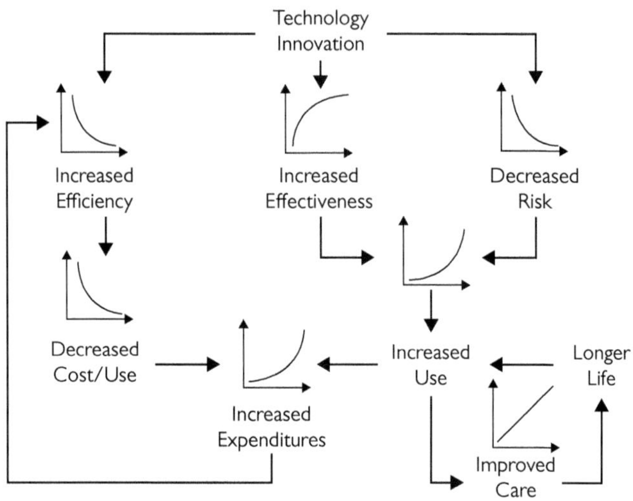

Fig 7.6 The dynamics of escalating healthcare costs.

Conclusions

The six case studies discussed in this chapter illustrate the broad range of applicability of computational modeling to designing, developing, and deploying the future health and well-being ecosystem. Big data and interactive visualization technology have enabled great advances. The Internet of Things and AI, in its several forms, will further advance the state of modeling and decision support.

It is easy to imagine evidence-based and intelligent support for clinicians, patients, and their families. Economic, political, and social fragmentation may be slow to fade, but model-based design will increasingly provide what users really need and desire. As has been seen with other digital technologies, the marketplace will drive adoption, despite lingering fragmentation. The next chapter considers some of the ingredients of such change.

Key Points

- Computational modeling can substantially contribute to exploring possible futures for health and well-being.
- Patient, provider, and payer data sets can be used to parameterize these computational models.

- Large interactive visualizations can enable a wide range of stakeholders to participate in exploring possible futures.
- Policy flight simulators can enable projecting likely impacts of policies, for example, alternative payment schemes, before they are deployed.
- There is enormous variety in healthcare, including patients, providers, and payers, as well as the economic and social circumstance in which they operate.
- Computational models can be invaluable for projecting the impacts of this variety and considering how system and policy designs should be tailored.

REFERENCES

Boustany, K., Pennock, M. J., Bell, T., Boustani, M., & Rouse, W. B. (2016). *Leveraging computer simulation models in healthcare delivery redesign*. Indianapolis, IN: Indiana University Health, Aging Brains Care Center.

CBO. (2008). *Technological change and the growth of health care spending*. Washington, DC: U.S. Congress, Congressional Budget Office.

Goldsmith, J., & Bajner, R. (2017, November 10; updated November 15). 5 ways U.S. hospitals can handle financial losses from Medicare patients. *Harvard Business Review*. Retrieved from http://hbr.org/2017/11/5-ways-u-s-hospitals-can-respond-to-medicares-mounting-costs.

LaMantia M. A., Alder C. A., Austrom M. G., Cottingham A. J., Litzelman D. K., Boustany K. C., & Boustani M. A. (2014). The Aging Brain Care Medical Home program: Accomplishments and lessons learned at one year. *Alzheimer's & Dementia: The Journal of the Alzheimer's Association*, *10*(4), P209.

Naylor, M. D. (2012). Advancing high value transitional care: The central role of nursing and its leadership. ***Nursing Administration Quarterly***, *36*(2), 115–126.

Naylor, M. D., Hirschman, K. B., Hanlon, A. L., Bowles, K. H., Bradway, C., McCauley, K. M., & Pauly, M. V. (2014). Comparison of evidence-based interventions on outcomes of hospitalized, cognitively impaired older adults. *Journal of Comparative Effectiveness Research*, *3*(3), 245–257.

Park, H., Clear, T., Rouse, W. B., Basole, R. C., Braunstein, M. L., Brigham, K. L., & Cunningham, L. (2012). Multi-level simulations of health delivery systems. A prospective tool for policy, strategy, planning and management. *Journal of Service Science*, *4*(3), 253–268.

Pennock, M. J.,Yu, Z., Hirschman, K. B., Pepe, K. P., Pauly, M. V., Naylor, M. D., & Rouse, W. B. (2018). Developing a policy flight simulator to facilitate the adoption of an evidence-based intervention. *IEEE Journal of Translational Engineering in Health and Medicine*, *6*(1), 1–12.

Rouse, W. B. (2015). *Modeling and visualization of complex systems and enterprises: Explorations of physical, human, economic, and social phenomena*. Hoboken, NJ: Wiley.

Rouse, W. B. (2016). *Universities as complex enterprises: How academia works, why it works these ways, and where the university enterprise is headed.* Hoboken, NJ: Wiley.

Rouse, W. B., & Cortese, D. A. (Eds.). (2010). *Engineering the system of healthcare delivery.* Amsterdam: IOS Press.

Rouse, W. B., Johns, M. M. E., & Pepe, K. (2017). Learning in the healthcare enterprise. *Journal of Learning Health Systems, 1*(4), e10024.

Rouse, W. B., Johns, M. M. E., & Pepe, K. (2019). Service supply chains for population health: Overcoming fragmentation of service delivery ecosystems. *Journal of Learning Health Systems.* 3 (2), https://doi.org/10.1002/lrh2.10186.

Rouse, W. B., Naylor, M. D., Yu, Z., Pennock, M. P., Hirschman, K. B., Pauly, M. V., & Pepe, K. P. (in press). Policy flight simulators: Accelerating decisions to adopt evidence-based health interventions. *Journal of Healthcare Management.*

Rouse, W. B., & Serban, N. (2014). *Understanding and managing the complexity of healthcare.* Cambridge, MA: MIT Press.

Yu, Z., Rouse, W. B., Serban, S., & Veral, E. (2016). A data-rich agent-based decision support model for hospital consolidation. *Journal of Enterprise Transformation, 6*(3/4), 136–161.

CHAPTER 8

Intelligent Systems

I have a folder on my backup disk with articles and summaries of articles I have read over the past 20 years or so in the process of writing articles and books. There are over 3,000 items in this folder. Over 100 of them relate to AI and have been added in the past 18 months. AI is a hot topic again—the third time since the 1950s. The hype is as great as ever. Nevertheless, big data, compute power, and visualization technology have enormously advanced. Ideas that were fanciful a few years ago now seem feasible.

The idea that intelligence can be automated, replacing millions of humans in routine jobs, has received an enormous amount of attention (e.g., Auerswald, 2017;Beyer, 2016; Brynjolfsson & McAfee, 2014). Various pundits have projected dramatic disruptions of the economy as robots, or their equivalent, pervasively provide an increasing range of services. There has been considerable debate about the extent to which completely "hands-off" automation will be possible and how legal issues will be addressed.

Undoubtedly, there are many jobs that involve 100 % routine, highly repeatable tasks that will become totally automated. There are many more jobs that are partly routine and partly nonroutine and will be amenable to automation that augments humans who are responsible for the nonroutine aspects of these jobs. This chapter addresses the ways in which human intelligence in such situations can be augmented rather than replaced.

First, consider several observations about contemporary AI. Machine learning—or deep learning—applications have demonstrated impressive capabilities to perform tasks such as recognizing pictures and speech, detecting anomalous behaviors, and other pattern-oriented functions. The neural network algorithms underlying machine learning are composed of multiple layers

Computing Possible Futures. William B. Rouse. Oxford University Press (2019). © William B. Rouse 2019.
DOI: 10.1093/oso/9780198846420.001.0001

involving both linear and nonlinear transformations. Conclusions reached by machine learning are, in general, not explainable, in the sense that the computational system cannot explain why it is making particular recommendations.

The implications are fairly clear. To the extent that decisions emanating from machine learning are always 100 % correct, then the action systems, human or otherwise, can simply execute the recommended decisions. If recommendations will occasionally be rejected, or should be rejected, then the lack of explanation capabilities will impose responsibilities on humans that will require decision support. This suggests the need for an intelligent interface layer between the machine learning capabilities and the action systems, particularly when human decision-makers are ultimately responsible for final decisions.

An intelligent interface needs to understand human decision-makers' intentions and provide support needed for successful pursuit of these intentions. Humans' intentions are very context dependent and change in time, depending on external circumstances and the intentions and actions of a range of stakeholders such as, for example, patients, customers, competitors, and, in the case of a driver, other drivers on the road. Consequently, an underlying time-varying workflow model is required that provides explicit representation of humans' goals, plans, scripts, and tasks, as well as information and control requirements. These notions come together in an approach to augmenting intelligence.

Succinctly, this chapter focuses on how intelligent systems technology can augment human behavior and performance, rather than replace it. I emphasize model-based decision support for a range of personnel—operators, designers, clinicians, and others. I also address the promise and perils associated with this technology.

AI and Machine Learning

One could argue that AI began with Ada Lovelace in the mid-1800s. However, many would agree that the field began in earnest with Alan Turing's landmark paper. His article on the Imitation Game unveiled his test for a machine's ability to exhibit intelligent behavior. It has remained an important philosophical construct with AI.

The emerging field of AI was recognized at the 1956 Dartmouth College AI Conference, which was led by John McCarthy, Marvin Minsky, Claude

Shannon, and Nathaniel Rochester. Marvin Minsky's PhD thesis led 15 years later to his book on perceptrons, which he co-wrote with Seymour Papert. Frank Rosenblatt's work on perceptrons appeared soon after Minsky's thesis. Allen Newell, John Shaw, and Herbert Simon published work on the General Problem Solver in 1959.

Even this early, differences of approaches were clear. Perceptrons were based on statistical methods for pattern recognition. This approach foreshadowed the success of the multiple layer networks of today's deep learning systems, which require tremendous computing power and data sets not available in those early days. Symbolic logic was adopted for problem-solving or reasoning. This approach presaged the rise of expert systems, and the challenges of manually building and maintaining large rule-based knowledge systems. Of course, using a network model to recognize an object is a rather different task from making a sequence of tests to troubleshoot an electronic circuit with a rule-based knowledge system.

In the 1960s, Joseph Weizenbaum introduced Eliza, which simulated "conversation" by matching patterns and substituting key words that gave users an illusion of understanding, despite the computer having no means for understanding the context of the dialog. A richer approach to language was Roger Schank's Conceptual Dependency Model, which eventually led to major contributions to natural language understanding.

The 1970s saw applications of AI to enhance medical diagnosis and treatment, starting perhaps with MYCIN. However, a report by James Lighthill criticized AI for articulating and then failing in its pursuit of grandiose objectives. This report, and other forces, led to the First AI Winter, with substantial DARPA funding cuts.

The 1980s saw the growth of expert systems, led by Edward Feigenbaum. These rule-based systems were built from "knowledge engineering" with subject matter experts. DARPA's Pilot's Associate Program emerged to leverage expert systems technology. Our basic research on intelligent interfaces (discussed below) was funded by a variety of agencies; this DARPA program provided the means to bring the pieces together.

The late 1980s saw the Second AI Winter. The Lisp machine market collapsed. Japan's Fifth Generation project fizzled. DARPA funding cuts happened again. This period saw computing move from Lisp machines, to Sun engineering workstations, to desktop PCs.

In the 1990s, there were several real applications. In 1991, the ISX Corporation created and deployed DART (Dynamic Analysis and Replanning Tool). DART was used by the U.S. military in the Middle East to optimize and schedule the transportation of supplies or personnel and solve other logistical problems. In 1997, IBM's Deep Blue defeated the chess master Garry Kasparov. Another major demonstration of IBM's capabilities came when Watson won *Jeopardy!* in 2011.

The 2000s also saw the maturation of deep learning, first at universities, and then at Microsoft, Google, and other companies. Deep learning is discussed in some depth in the next section. It is nevertheless worth noting here a trend over the 60+ years since Turing. Early innovations were associated with people, often individuals at universities. Universities helped create large open data sets and competitions that led to measurable progress and accumulation of knowledge. Teams, often working at large companies, have accomplished later innovations, often with far more computing resources.

Another trend is also of note. Early research focused on two dominant approaches: statistically based learning for pattern recognition, and rule-based or symbolic logic for problem-solving and reasoning. Big data and almost-free computing power have allowed enormous advances in machine-based pattern recognition. I argue later that both are still needed to augment intelligence.

Contemporary AI

AI based on symbolic logic worked where rules and definitions were very clear, such as, for example, in domains such as mathematics and chess. However, the symbolic logic approach was overwhelmed by many pattern recognition tasks. AI based on layered neural nets, now termed "deep machine learning," has been successful for speech recognition, image recognition, language translation, and driverless cars. Each subsequent layer looks for patterns in the previous level's patterns. Early on, this approach was called connectionism or distributed parallel processing.

Deep learning works well when trained with large numbers of examples, but this is not feasible for many tasks such as, for example, reasoning about and solving novel problems. Further, as the Stanford One Hundred Year Study on Artificial Intelligence notes, "No machines with self-sustaining long-term goals and intent have been developed, nor are they likely to be developed in the near future."

This limitation is well illustrated by a recent experience. A couple of weeks ago, I participated in an advisory board meeting where a presentation was

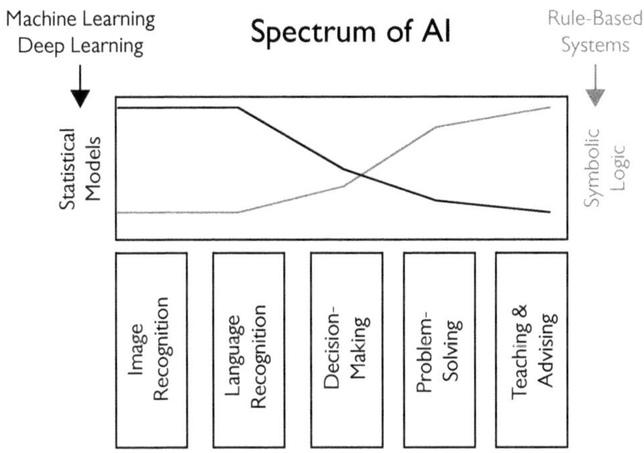

Fig 8.1 The spectrum of AI.

made on the state of machine learning in the military. The speaker explained how the machine learning had been trained to identify different classes of Chinese military vessels by using aerial reconnaissance data as input. He showed how it had recently correctly identified each vessel in the South China Sea, noting that it performed better than human classification using the same imagery.

I asked if he could tell me if the deployed armada intended to attack Taiwan. He said, "No." However, he added that if I could provide pictures of each of the last 10,000 times that China had attacked Taiwan, he could train the machine learning to do this. The key point here is that machine learning is much better for recognition and classification tasks than for prediction tasks.

Figure 8.1 portrays the spectrum of AI from statistical models to symbolic logic. Machine learning can be superior for recognition and classification. Symbolic logic is better for problem-solving, teaching, and advising. Decision-making can benefit from both approaches, as I later illustrate. An example will help make this point. Consider a system intended to teach algebra. I can imagine training machine learning with many thousands of problems and solutions so that it can readily solve such problems.

However, if the purpose of the system were to teach humans how to do algebra, this system would be useless, because it could not explain how it was solving problems. In fact it is doubtful that humans could employ the machine's approach even if it could explain it. This reminds me of a recent article that reported on how many instances of stop signs were needed to train machine learning to recognize these signs. In contrast, when my three-year old grandson asked, "What is that sign called?" and I told him it was a stop sign, he insisted on pointing out stop signs for the rest of the day. He only needed one example.

Elements of Intelligent Support

The foregoing sets the stage for the main theme of this chapter. In many situations, AI will be used to augment human intelligence, rather than being deployed to automate intelligence and replace humans. In Chapter 6, I addressed augmenting human intelligence by increasing humans' potential to perform via training. In this chapter, I discuss aiding.

Aiding is concerned with directly augmenting human behavior and performance. We can use models such as those depicted in Chapter 6 to understand human behavior and performance—in real time as it is happening. This understanding can be used by the aiding subsystem to assist humans by tailoring their displays and controls, monitoring behaviors, and adapting assistance according to humans' needs and desires (Rouse, 2007, 2015; Rouse & Spohrer, 2018).

Before discussing functions needed to augment intelligence, it is useful to consider how we go about designing aiding systems. We start by understanding the tasks to be aided and the desired user experience (often referred to as "UX") when aided. This is often accomplished by drafting scenarios—stories—of users performing the targeted tasks with the aiding assisting them. We iterate until the scenarios are desirable and, hopefully, compelling.

With the user experience defined, we shift attention to the user interface (often referred to as "UI") in terms of displays and controls—what the users see and what they can do. Once the user-interface design is satisfactory, we shift attention to the technical solution for the functionality of the aiding. This approach is in contrast with a more typical approach that starts with the technical solution. Once that is working, they slap on a user interface and deploy it to users. It is common for users to find the solution opaque and confusing, sometimes rejecting what was potentially a good idea.

Information Management

Several functions are central to intelligent support. One function is information management (Rouse, 2007). This involves information selection (what to present) and scheduling (when to present it). Information modality selection involves choosing among visual, auditory, and tactile channels. Information formatting concerns choosing the best levels of abstraction (concept) and aggregation (detail) for the tasks at hand. AI can be used to make all these choices in real time as the human is pursuing the tasks of interest.

Intent Inferencing

Another function is intent inferencing (Rouse, 2007). Information management can be more helpful if it knows both what humans are doing and what they intend to do. Representing humans' task structure in terms of goals, plans, and scripts (Schank & Abelson, 1977) can enable making such inferences—see Figure 8.2. Scripts are sequences of actions to which are connected information and control requirements. When the intelligence infers what you intend to do, it then knows what information you need and what controls you want it to execute.

One of the reasons that humans are often included in systems is because they can deal with ambiguity and figure out what to do. Occasionally, what they decide to do has potentially unfortunate consequences. In such cases, "human errors" are reported. Errors in themselves are not the problem. The consequences are the problem.

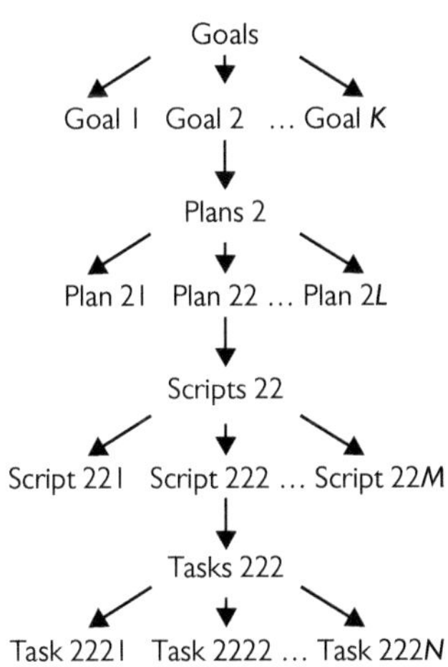

Fig 8.2 Structure of a goal-plan graph.

Error-Tolerant Interfaces

For this reason, another function is an error-tolerant interface (Rouse, 2007). This requires capabilities to identify and classify errors, which are defined as actions that do not make sense (commissions), or the lack of actions that seem warranted at the time (omissions). Identification and classification lead to remediation. This occurs at three levels: monitoring, feedback, and control. Monitoring involves collecting more evidence to support the error assessment. Feedback involves making sure the humans realize what they just did. This usually results in humans immediately correcting their errors. Control involves the automation taking over, for example, applying the brakes, to avoid the imminent consequences.

Adaptive Aiding

The notion of taking control raises the overall issue of whether humans or computers should perform particular tasks. There are many cases where the answer is situation dependent. Thus, this function is termed adaptive aiding (Rouse, 2007). The overall concept is to have mechanisms that enable real-time determination of who should be in control. Such mechanisms have been researched extensively, resulting in a framework for design that includes principles of adaptation and principles of interaction. I have proposed a First Law of Adaptive Aiding—*computers can take tasks, but they cannot give them.*

Intelligent Tutoring

Another function is intelligent tutoring to both train humans and keep them sufficiently in the loop to enable successful human task performance when needed. Training usually addresses two questions: (1) how the system works, and (2) how to work the system. Keeping humans in the loop addresses maintaining competence. Unless tasks can be automated to perfection, humans' competencies need to be maintained. Not surprisingly, this often results in training versus aiding trade-offs, for which guidance has been developed (Rouse, 2007).

Overall Architecture

Figure 8.3 provides an overall architecture for augmenting intelligence. The intelligent interface, summarized above, becomes a component in this broader concept. The overall logic is as follows:

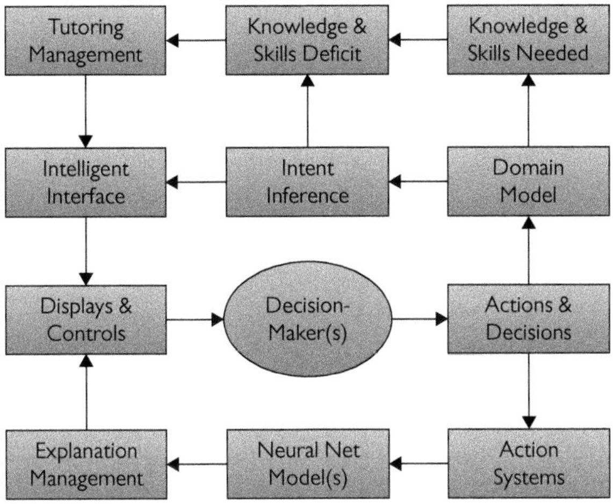

Fig 8.3 Overall architecture of augmented intelligence.

- Humans see displays and controls and then decide and act. Humans need not be concerned with other than these three elements of the architecture. The overall system frames humans' roles and tasks and provides support accordingly.
- The intent inference function infers what task(s) humans intend to do. This function retrieves information and control needs for these task(s). The information management function determines displays and controls appropriate for meeting information and control needs.
- The intelligent tutoring function infers humans' knowledge and skill deficits relative to these task(s). If humans cannot perform the task(s) acceptably, the information management function either provides just-in-time training or informs adaptive aiding (see below) of the humans' need for aiding.
- Deep learning neural nets provide recommended actions and decisions. The explanation management function provides explanations of these recommendations to the extent that explanations are requested. This function is elaborated below.
- The adaptive aiding function, within the intelligent interface, determines the human's role in execution. This can range from manual to automatic control, with execution typically involving somewhere between these extremes. The error monitoring function, within the intelligent interface, detects, classifies, and remediates anomalies.

Note that these functions influence each other. For example, if adaptive aiding determines that humans should perform a task, intelligent tutoring assesses the availability of the necessary knowledge and skills and determines the training interventions needed, and information management provides the tutoring experiences to augment knowledge and skills. On the other hand, if adaptive aiding determines that automation should perform a task, intelligent tutoring assesses humans' abilities to monitor automation, assuming such monitoring is needed.

Explanation Management

As discussed in the section "Contemporary AI," neural network models cannot explain their (recommended) decisions. This would seem to be a fundamental limitation. However, science has long addressed the need to understand systems that cannot explain their own behaviors. Experimental methods are used to develop statistical models of input–output relationships. Applying these methods to neural network models can yield mathematical models that enable explaining the (recommended) decisions, as shown in Figure 8.4.

Given a set of independent variables *X*, a statistical experiment can be designed, for example, a fractional factorial design, that determines the combinations of values of *X* to be input into the neural net model(s). These models, typically multilayered, have "learned" from exposure to massive data lakes with labeled instances of true positives, and possibly false positives and false negatives. True negatives are the remaining instances.

The neural net models yield decisions, *D*, in response to the designed combinations of *X*. A model, *D*(*X*), is then fit to these input–output data sets. Explanation generation then yields explanations, *E*(*D*), that are based on the

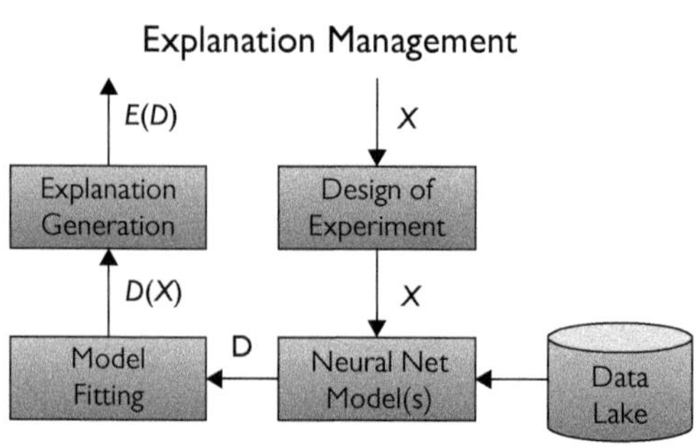

Fig 8.4 Explanation management function.

attributes and weights in the fitted model. The result is a first-order, that is, non-deep, explanation of the neural net (recommended) decisions.

As noted earlier, the paradigm underlying the approach shown in Figure 8.4 is the standard paradigm of empirical natural science. Thus, it is clear it will work, that is, yield rule-based explanations, but will it be sufficient to help decision-makers understand and accept what the machine learning recommends? I imagine this will depend on the application.

As an example, consider control theory. Optimal stochastic control theory includes both optimal estimation and optimal control. Determining the optimal solution across both estimation and control involves rather sophisticated mathematics. We could apply the method shown in Figure 8.4 to the optimal control actions resulting from the solution of this stochastic control problem.

We would not be able to infer the nature of the underlying sophisticated mathematics. Instead, we would likely unearth something akin to classic PID controllers, where the acronym stands for proportional, integral, and derivative attributes of the errors between desired and actual states. It has been shown that this provides a reasonable explanation of optimal control actions.

Learning Loops

Figures 8.3 and 8.4 include both explicit and implicit learning loops. The statistical machine learning loop will be continually refining the relationships in its layers, either by supervised learning or by reinforcement learning. This will involve balancing exploration (of uncharted territory) and exploitation (of current knowledge). This may involve human designers and experimenters not included in Figures 8.3 and 8.4. Of particular interest is how machine learning will forget older data and examples that are no longer relevant, for example, a health treatment that has more recently been shown to be ineffective.

The rule-based learning loops in Figures 8.3 and 8.4 are concerned with inferring rule-based explanations of the recommendations resulting from machine learning (Figure 8.4) and inferring human decision-makers' intentions and state of knowledge (Figure 8.3). Further, learning by decision-makers is facilitated by the tutoring function in Figure 8.3.

Thus, the AI will be learning about phenomena, cues, decisions, actions, and so on in the overall task environment. The decision-makers will learn about what the AI is learning, which is expressed in more readily understandable rule-based forms. The intelligent support system will be learning about the decision-makers' intentions, information needs, and so on, as well as influencing what the decision-makers learn.

Paradigms Employed

Considering the example modeling capabilities discussed in this section in terms of the representational paradigms discussed in Chapter 2, five are represented:

- control theory: adaptive aiding involves the control of queues of tasks and dynamic allocation of tasks to humans or computers;
- queuing theory: the model represents the queues of tasks to be performed either by humans or by computers;
- network theory: goal-plan graphs represent relationships between goals, plan scripts, tasks, and information and control requirements;
- estimation theory: statistical models are used to support explanation management; and
- problem-solving theory: rule-based models are employed to manage all the functions within the intelligent interface.

Thus, five of the eight paradigms from Chapter 2 are of use in the intelligent interface. The instantiation of these paradigms depends, of course, on the context of the applications.

Case Studies

Many of the earlier research and applications of the notions elaborated in this section focused on operation and maintenance of complex engineered systems such as aircraft, power plants, and factories. The tasks associated with such systems are usually well understood. One application focused on electronic checklists for aircraft pilots (Rouse, 2007). The results were sufficiently compelling to motivate inclusion of some of the functionality on the Boeing 777 aircraft.

The overall conceptual architecture of Figure 8.3 has been developed and applied several times to tasks that are sufficiently structured to be able to make the inferences needed to support the functionality outlined here (Rouse, 2007). The explanation management function of Figure 8.4 was added more recently to take advantage of recent developments in machine learning (Rouse & Spohrer, 2018).

Process Control

We conducted a series of studies of human detection, diagnosis, and compensation for systems failures in a simulated environment named PLANT—Production Levels and Network Troubleshooting. We focused primarily on what operators need to know to perform well. We developed rule-based models of each operator by using a framework called KARL—Knowledgeable Application of Rule-Based Logic—building on the constructs in Table 6.1 of Chapter 6 (Rouse, 2007).

The idea emerged to use the models of each operator as an assistant to each operator. In other words, each person's assistant was a computational model of himself or herself. Thus, the "second opinion" provided was very much biased to agree with the operator's own opinion. In general, this assistant improved detection, diagnosis, and compensation in terms of faster responses and fewer errors.

One of the failures of PLANT was a safety system failure. The safety system was designed to trip (shut off) pumps and valves when fluid levels became too high or low. The safety system failure resulted in random trips unrelated to fluid levels. Operators were told about this possibility, but their training did not include experiencing it.

None of the operators without the computational assistant were successful in dealing with this one-time failure. They were rather confused by it. Most of the operators with the assistant correctly diagnosed this failure and shut the safety system off. Their computational assistant had no knowledge of the possibility of a safety system failure. In fact, its consistent advice was that nothing was wrong.

The operators reacted to this advice by saying something like "No, everything is not fine. Something unusual has happened." They then correctly diagnosed the safety system failure.

The key point here is that bad advice improved performance. This suggests that there are subtleties to the impacts of intelligent assistance. Augmented intelligence can result even when the computer is a flawed assistant. This phenomenon merits much more research, but this result does emphasize the possibility of unexpected outcomes.

Pilot's Associate

My research on aiding operators and maintainers had been ongoing for 15 years when Lockheed Martin contacted me. They had tried to contact me at the University of Illinois, but I had moved to Georgia Tech a few years earlier. This was fortunate, because their facility was now only a few miles away, rather than over 600.

Lockheed was intending to bid on the Pilot's Associate Program sponsored by DARPA. They wanted to know if our aiding concepts, for example, information management, error-tolerant interfaces, and adaptive aiding, could be incorporated into their concept of an artificially intelligent copilot for fighter aircraft. We developed a concept similar to the one shown in Figure 8.3 to enable an intelligent pilot–vehicle interface (Rouse, 2007).

This was the largest project our company, Search Technology, had ever undertaken. Working with Lockheed and several other subcontractors, we developed prototypes of our functionality, integrated it with others' functionality, and evaluated an ongoing series of integrated prototypes. One issue that received significant attention was how best to validate the intelligence of the Pilot's Associate, particularly the circumstances under which it would provide bad advice (Rouse, 2007).

Designer's Associate

Soon after winning the Pilot's Associate project, we won a contract with the U.S. Air Force to develop a Designer's Associate to provide intelligent assistance to aircraft cockpit designers. This was also a large contract, and the two projects combined led to strong company growth. It was a challenge to recruit people with the right competencies, as is now the case for AI and data analytics.

We needed to know a lot more about cockpit design. We interviewed designers at the seven aerospace companies then designing military aircraft—there are only three now. We found what can best be characterized as a complex sociotechnical system, laced with technological, financial, market, and political issues. Designing a cockpit involves much more complexity and many more trade-offs than taking off or landing an airplane.

We used these findings to develop scenarios of designers interacting with the Designer's Associate to formulate and address cockpit design issues. Designers were asked to review and evaluate these scenarios. One question was "To what extent does the scenario portray how you do your job?" One response had a profound impact. The designer, George, said, "Not at all, but I sure wish it did."

With this response, we realized that our solution had to go far beyond just supporting current work practices. The Designer's Associate needed to enable transformation of designers' work into the digital environment that has become common in the aerospace domain. We developed a conceptual design for the Designer's Associate along the lines of that shown in Figure 8.3 (Rouse, 2007, 2015) and then used it to evaluate a range of prototypes.

We came to think that the Designer's Associate should not be a monolithic intelligent system that supports all design decision-making. Instead, it should be a suite of complementary tools, which share data and provide intelligent assistance in the context of each tool. The result was the *Advisor Series* of intelligent, computational tools that was discussed in Chapter 4.

Advisor Series

The *Situation Assessment Advisor*, based on *Start Where You Are* (Rouse, 1996), and the assessment tool provided with *Don't Jump to Solutions* (Rouse, 1998) were rule-based expert systems. The rules for assessing market situations or organizational delusions were gleaned from extensive reviews of the literature, augmented with knowledge from our work with over 100 companies. I find it difficult to imagine how we would now do this with machine learning, as there are not thousands of documented examples of the 100 combinations of current and future market situations that we used, nor of the many variations of the 13 organizational delusions described in the tools.

For both tools, management teams would answer a lengthy set of questions. For assessing their situation, the questions were about their markets, their positions in these markets, and their current and projected financial performance. For assessing organizational delusions, the questions related to their beliefs about their customers, offerings, organization, and competitors. Given their answers to these questions, the expert systems in the respective tools would display an assessment for the management team to discuss.

One of the best-liked features of these tools was their ability to present examples of what other companies with these assessments did and the extent to which these efforts succeeded or failed. Having 10 or so examples of how other companies pursue change prompted much creative discussion by the teams. In fact, the assessments were, to an extent, just abstractions that enabled retrieving the kinds of examples that would motivate teams. As indicated in Chapter 6, one of the most important roles of our tools has been to enable and focus the creativity of management teams. This is a very important aspect of augmented intelligence.

Cognitive Assistants

Alexa and Siri have been touted as cognitive assistants. They do provide some value in terms of answering straightforward questions, turning on or off the lights, and increasing or lowering the TV volume. However, this is a very limited view of cognitive assistance.

We have been focusing on cognitively assisting work, ranging from professionals such as clinicians or engineers, to disabled and older adults. Rouse, Johns, and Pepe (2019) consider cognitive assistance for clinicians and patients. Rouse and McBride (2019) discuss assistive technologies for disabled and older adults.

What does a cognitive assistant need to know and be able to do? It needs two broad classes of knowledge. First, it needs knowledge of the domain of the user, that is, his or her work, whether it is employment or the activities of daily life. Second, it needs to understand the particular user in terms of calendars, contacts, preferences, and relationships in general.

Considering what a cognitive assistant needs to be able to do, it needs to be able to interact using natural language, make inferences of the user's intentions, support the pursuit of these intentions, and continually learn about the user and their domain. This is far too much to expect for an off-the-shelf Alexa or Siri, but it is entirely feasible for specific work and particular users. I am sure this will be the emergent reality, although I am less sure of the time frame for emergence.

Promises, Perils, and Prospects

This chapter began by citing various pundits regarding what can be automated, which could replace millions of humans in routine jobs (e.g., Auerswald, 2017; Beyer, 2016; Brynjolfsson & McAfee, 2014). Their prognostications suggest both promises and perils from AI and machine learning, and from data analytics more broadly.

The promise is epitomized by the prospects for evidence-based decision-making for health, education, social policy, and so on (Abraham & Haskins, 2017; Haskins & Margolis, 2014). Enormous data sets, processed by both traditional analytics and machine learning, will enable learning what works and what does not. Of course, there are perils.

Lazer and colleagues (2014) suggest that there are two factors that may result in our being misled. First, big data hubris leads to assumptions that big data can substitute for traditional data collection. Second, algorithm dynamics results in the constant changing of algorithms and data-gathering practices, such that any notions of baselines disappear. Thus, we cannot necessarily trust the data, and how it is processed continually changes.

Of course, Amazon, Apple, Facebook, and Google have already shown what can be done with large data sets, and that this is not always to people's benefit. O'Neil (2016) characterizes these capabilities as "weapons of math destruction." She argues that these weapons increase inequality and threaten democracy, due to those with high-end information infrastructures disproportionately benefiting, as well as through the manipulation of electoral processes.

Harris (2017) reports on the reproducibility crisis in biomedical research. Due to sloppy practices and the rush to publish, the majority of results reported in the biomedical literature cannot be reproduced when these studies are redone. Thus, the evidence base meant to inform medical practice is highly flawed. A central issue is the incentive and reward system in biomedical research, for example, promotion and tenure depends on publishing a steady stream of articles in top journals.

Finally, cybersecurity is a pervasive concern, with frequent reports of large-scale hacking. Personal information is either stolen or provided to third parties without the owners of this information having provided permission. All of the above factors lead to issues of trust. Table 8.1 provides a perspective on trust.

Trust depends on humans' expectations of cognitive assistants and the experienced performance of these assistants in terms of quality of recommendations and possibly execution. This suggests that the relationships between users and cognitive assistants will play a central role in determining the extent to which cognitive assistants are seen as beneficial and are used.

Beyond the promises and perils, there are prospects for intelligent systems technologies that are likely to be adopted and deployed. Sampath and Khargonekar (2018) propose four levels of automation:

- L3: socially responsible automation—socioeconomic impact;
- L2: human-centered automation—workforce development and enrichment;
- L1: Performance-driven automation—focus on productivity, quality, accuracy, and speed; and
- L0: cost-focused automation—focus on labor reduction.

The lowest level is focused on reducing labor costs. This motivation has driven many investment decisions. The next level is concerned with the productivity, quality, accuracy, and speed of the overall human–machine system. The Pilot's Associate discussed earlier is a good example of this type of automation.

Table 8.1 ***How humans may demonstrate trust***

	Cognitive Assistant Performance (Relative to Expectations)			
Human's Level of Trust	Demonstrated Long-Term Good Performance	Demonstrated Near-Term Good Performance	Low Frequency of Poor Performance	High Frequency of Poor Performance
Human Allows Assistant to Perform Its Chosen Actions	Yes	Maybe	No	No
Human Performs Actions Assistant Recommends	Yes	Yes	Maybe	No
Human Takes into Account Assistant's Recommendations	Yes	Yes	Maybe	No
Human Ignores Assistant, Perhaps Turning It Off	No	No	Maybe	Yes

The next level, human-centered automation, addresses workforce development and enrichment. The Designer's Associate is an example of this type of automation in that its goal was to transform designers' jobs and enable better performance. The *Advisor Series* of tools fits here are well.

The highest level, socially responsible automation, addresses socioeconomic impacts. Our work on assistive technologies for disabled and older adults belongs here, with its emphasis on the fulfillment of work and the economic empowerment of this rapidly growing proportion of our population. Whether the work is employment or the activities of daily life, assistive technologies have the potential to transform lives.

Conclusions

This chapter has addressed intelligent systems, with an emphasis on augmenting human intelligence. After providing a brief history of AI and machine learning, I focused on the type of problems for which each approach to AI is most appropriate.

I next addressed the elements of intelligent support to enable augmented intelligence. This was followed by elaboration of an architecture for intelligent support, including explanation management that can enable the integration of elements of AI based on symbolic logic and machine learning.

A wide range of case studies was reviewed, ranging from process control, to aircraft piloting, to aircraft design, to various intelligent tools and cognitive assistants. Finally, I addressed the promises, perils, and prospects for intelligent system technologies.

Key Points

- Design of aids should begin with defining the user experience, proceed to designing the user interface to support this experience, and then focus on the enabling technologies.
- Intelligent aids should be considered for enhancing human performance; the extent of success will depend on the domain of application and the potential performance of the aid.
- Intelligent aids are inherently model based, drawing upon symbolic logic, mathematical paradigms, and/or statistical models; understanding the

underlying modeling assumptions is key to establishing confidence in and trust of such aids.

- Intelligent systems technology has much promise, but also many perils that warrant attention; its prospects depend on well-reasoned strategies for development and adoption.

REFERENCES

Abraham, K. G., & Haskins, R. (2017). *The promise of evidence-based policymaking.* Washington, DC: Commission on Evidence-Based Policy Making.

Auerswald, P. E. (2017). *The code economy: A forty-thousand-year history.* New York: Oxford University Press.

Beyer, D. (Ed.). (2016). *The future of machine intelligence: Perspectives from leading practitioners.* Sebastopol, CA: O'Reilly Media.

Brynjolfsson, E., & McAfee, A. (2014). *The second machine age: Work progress, and prosperity in a time of brilliant technologies.* New York: Norton.

Harris, R. (2017). *Rigor mortis: How sloppy science creates worthless cures, crushes hope, and wastes billions.* New York: Basic Books.

Haskins, R., & Margolis, G. (2014). *Show me the evidence: Obama's fight for rigor and results in social policy.* Washington, DC: Brookings Institution.

Lazer, D., Kennedy, R., King, G., & Vespignani, A. (2014). The parable of Google flu: Traps in big data analysis. ***Science***, *343*(14), 1203–1205.

O'Neil, C. (2016). *Weapons of math destruction: How big data increases inequality and threatens democracy.* New York: Broadway Books.

Rouse, W. B. (1998). *Don't jump to solutions: Thirteen delusions that undermine strategic thinking.* San Francisco, CA: Jossey-Bass.

Rouse, W. B. (2015). *Modeling and visualization of complex systems and enterprises: Explorations of physical, human, economic, and social phenomena.* New York: Wiley.

Rouse, W. B. (2007). *People and organizations: Explorations of human-centered design.* New York: Wiley.

Rouse, W. B. (1996). *Start where you are: Matching your strategy to your marketplace.* San Francisco, CA: Jossey-Bass.

Rouse, W. B., & McBride, D. K. (2019). A systems approach to assistive technologies for disabled and older adults. *The Bridge*, *49*(1), 32–38.

Rouse, W. B., & Spohrer, J. C. (2018). Automating versus augmenting intelligence. *Journal of Enterprise Transformation*. doi:10.1080/19488289.2018.1424059.

Rouse, W. B., Johns, M. M. E., & Pepe, K. (2019). Service supply chains for population health: Overcoming fragmentation of service delivery ecosystems. *Journal of Learning Health Systems*. doi:10.1002/lrh2.10186.

Sampath, M., & Khargonekar, P. P. (2018). Socially responsible automation: A framework for shaping the future. *The Bridge*, *48*(4), 45–52.

Schank, R., & Abelson, R. P. (1977). *Scripts, plans, goals and understanding: An inquiry into human knowledge structures.* Hillsdale, NJ: Erlbaum.

CHAPTER 9

Enterprise Transformation

In Chapter 4, I considered markets and competitors in terms of product and service offerings. Chapter 5 addressed technology adoption to enhance these offerings and gain competitive advantages. Chapter 8 discussed intelligent systems technologies and how they can augment human intelligence, potentially adding new dimensions to competitive advantage.

Successful adoption of the model-based approaches in these chapters can be helped or hindered by the nature of the enterprise. Companies such as Kodak, Polaroid, Xerox, Motorola, and Nokia developed technologies that were potential market innovations. However, these technologies remained on the shelf while these companies tried to continue milking the cash cows of their existing offerings. Competitors beat them to the marketplace, and these companies are now mere shadows of their former selves.

It is very difficult to successfully innovate when new offerings obsolete your current offerings. It can require transformation of your enterprise. However, it has been suggested that transforming an enterprise is akin to rewiring a building while the power is on. How can we design and develop a transformed enterprise while also avoiding operational disruptions and unintended consequences in the process? To address this question, we need a deeper understanding of the notion of enterprise transformation.

Our earlier studies (Rouse, 2005, 2006) have led us to formulate a qualitative theory: "Enterprise transformation is driven by experienced and/or anticipated value deficiencies that result in significantly redesigned and/or new work processes as determined by management's decision-making abilities, limitations, and inclinations, all in the context of the social networks of management in particular and the enterprise in general."

Computing Possible Futures. William B. Rouse. Oxford University Press (2019). © William B. Rouse 2019.
DOI: 10.1093/oso/9780198846420.001.0001

Context of Transformation

Enterprise transformation occurs in—and is at least partially driven by—the external context of the economy and markets. As shown in Figure 9.1, the economy affects markets that, in turn, affect enterprises. Of course, it is not quite as crisply hierarchical as indicated, in that the economy can directly affect enterprises, for example, via regulation and taxation. The key point is that the nature and extent of transformation are context dependent.

For public sector enterprises, the term "constituency" can replace the term "market." The financially oriented metrics shown in Figure 9.1 also have to be changed to reflect battles won, diseases cured, and so on. I will occasionally draw parallels between private and public sector enterprises; however, full treatment of these parallels is beyond the scope of this chapter.

There is also an internal context of transformation—the "intraprise" in Figure 9.1. Work assignments are pursued via work processes and yield work products, incurring costs. Values and culture, reward and recognition systems, individual and team competencies, and leadership are woven throughout the intraprise. These factors usually have strong impacts on an enterprise's inclinations and abilities to pursue transformation.

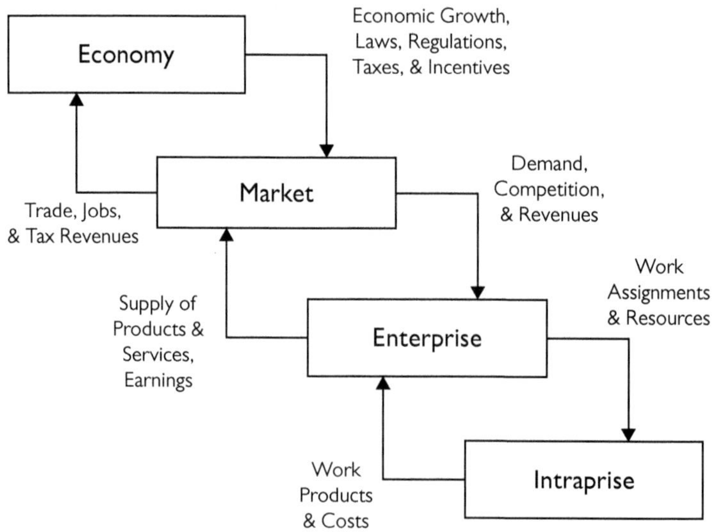

Fig 9.1 Context of enterprise transformation.

Modeling the Enterprise

Enterprise transformation occurs in the external context of Figure 9.1. The enterprise, with its internal strengths and weaknesses, and external opportunities and threats, operates within this broader external context. Possibilities for transformation are defined by the relationships between the enterprise and this context. The model of the enterprise as a system shown in Figure 9.2 provides a basis for understanding these possibilities.

Relationships among the elements of the enterprise system are as follows. Inputs affect both work processes and the enterprise state. For example, input resources (e.g., people, technology, and investment) affect both how work is done and how well it is done. As another example, input market conditions (e.g., demand and competition) affect the quality and pricing of products and services.

The concept of "state" is central to the theory of enterprise transformation. As discussed in Chapter 2, the state of a system is the set of variables and their values that enable assessing where the system is and projecting where it is going. We tend to think that financial statements define the state of an enterprise as a system. However, financial variables are usually insufficient to project the future of an enterprise, and a deeper characterization of state is needed. The "Balanced Scorecard" (Kaplan & Norton, 1996) or, deeper yet, an enterprise-oriented version of the "House of Quality" (Hauser & Clausing, 1988) are two possibilities.

Output is derived from the evolving state of the enterprise. For example, revenues can be determined from the numbers of units of products or services sold and the prices of these offerings. Determining profits requires also

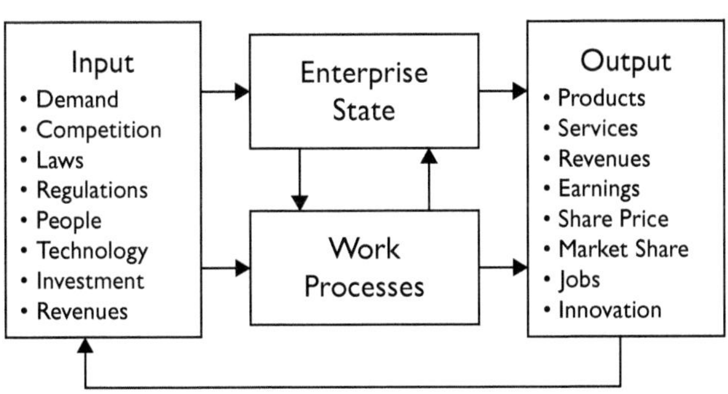

Fig 9.2 Elements of an enterprise system.

knowing the costs of providing offerings. Units sold relates, at least in part, to customer satisfaction as determined by product and service functionality, quality, and price, all relative to competing offerings.

The construct of "value" is central to the arguments that follow. The value of the enterprise is traditionally viewed as its market capitalization, that is, share price times number of outstanding shares. Share price is traditionally conceptualized as the NPV of future enterprise free cash flows, that is, revenues minus costs. This view of value is often characterized as "shareholder value."

From this perspective, state variables such as revenues, costs, quality, and price determine value. These variables are themselves determined by both work processes and architectural relationships among processes. Inputs such as investments of resources affect work processes. Coming full circle, the value of projected outputs influences how input resources are attracted and allocated.

Table 9.1 summarizes several examples of enterprise domains, processes, states, work, and value. It is important to note that value, for example, in terms of unit prices, will depend on the competing offerings from other enterprises. Similarly, the importance of any set of military objectives secured depends on the objectives secured by adversaries. Thus, as noted earlier, knowledge of context is essential to understanding enterprises as systems.

The examples in Table 9.1 serve to illustrate the multifaceted nature of value. It could be argued that all of the facets shown in the far-right column are simply intermediate surrogates for shareholder value; hence, shareholder value is the central construct. On the other hand, it is very difficult to argue that shareholder value, as traditionally defined, is the sole driver of enterprise

Table 9.1 ***Example domains, processes, states, work, and value***

Domain	Process	State	Work	Value
Manufacturing	Production	Work in process	Products	Unit price minus cost
Service	Delivery	People in queues	Transactions	Customer satisfaction
R & D	Research	Studies in progress	Technology options	Potential of options
Military	Operations	Positions of forces	Objectives secured	Importance of objectives

transformation. For many types of enterprises, shareholder value is the ultimate measure of success, but other forces such as markets, technologies, and the economy often drive change. Examples discussed later illustrate these forces.

Many fundamental changes address value from the perspective of customers and, to a much lesser extent, suppliers and employees. According to Peter Drucker (2001), "The purpose of a business is to create a customer." Thus, for example, while loss of market share and subsequent decreasing stock market valuation can be viewed as end effects in themselves, they also may be seen as symptoms of declining value of products and services as perceived by customers. Clearly, a broader view of value is needed (Slywotsky, 1996; Slywotsky & Morrison, 1997).

Qualitative Theory

Succinctly, experienced or expected value deficiencies drive enterprise transformation initiatives. Deficiencies are defined relative to both current enterprise states and expected states. Expectations may be based on extrapolation of past enterprise states. They may also be based on perceived opportunities to pursue expanded markets, new constituencies, technologies, and so on. Thus, deficiencies may be perceived for both reactive and proactive reasons.

Transformation initiatives involve addressing what work is undertaken by the enterprise and how this work is accomplished. The work of the enterprise ultimately affects the state of the enterprise, which is reflected, in part, in the enterprise's financial statements, Balanced Scorecard assessment, or the equivalent. Other important elements of the enterprise state might include market advantage, brand image, employee and customer satisfaction, and so on. In general, the state of the enterprise does not include variables internal to work processes.

This is due to the fact that we only need state estimates sufficient to enable explaining, predicting, and/or controlling future states of the system. To illustrate, the state of an aircraft is usually defined in terms of its location, speed, attitude, and so on, but not the current rpm of its fuel pumps, the air flow in its cabin, or the electron charge of its LED displays. Similarly, the state of an enterprise does not include the current locations of all its salespeople, the ambient temperatures in each of its factories, the water flow in the rest rooms, and so on.

Were we not able to define state at a higher level of aggregation and abstraction, the complexity of modeling airplanes or enterprises would be intractable.

Value Deficiencies Drive Transformation

More specifically, enterprise transformation is driven by perceived value deficiencies relative to needs and/or expectations due to

- experienced or expected downside losses of value, for example, declining enterprise revenues and/or profits;
- experienced or expected failures to meet projected or promised upside gains of value, for example, failures to achieve anticipated enterprise growth; or
- desires to achieve new levels of value, for example, via the exploitation of market and/or technological opportunities.

In all of these cases, there are often beliefs that change will enable remediation of such value deficiencies. Change can range from business process improvement to more fundamental enterprise transformation.

Work Processes Enable Transformation

In general, there are three broad ways to approach value deficiencies, all of which involve consideration of the work of the enterprise:

- improve how the work is currently performed, for example, reduce variability;
- perform the current work differently, for example, switch to web-enabled customer service; and
- perform different work, for example, outsource manufacturing and focus on service.

The first choice is basically business process improvement. This choice is less likely to be transformative than the other two choices. The second choice often involves operational changes that can be transformative depending on the scope of changes. The third choice is most likely to result in transforming the enterprise. This depends, however, on how resources are redeployed. Liquidation, in itself, is not necessarily transformative.

The need to focus on work processes is well recognized (e.g., Hammer & Champy, 1993; Womack & Jones, 1996). Re-engineered and lean processes have been goals in many transformative initiatives. Indeed, a focus on processes may, at least initially, require transformation of management's thinking about an enterprise. The extent to which this subsequently transforms the enterprise depends on the extent of changes and success in their implementation.

Transformation can also involve relationships among processes, not just individual work processes in and of themselves. These relationships are often framed in terms of an "architecture." It is common to express architectures in terms of multiple "views." The operational view is a description of the activities, operational elements, and information flows required to support enterprise operations. The technical view is a set of rules defining the interactions and interdependencies of system elements to assure compatibility and satisfaction of requirements. The system view describes the physical connections, locations, key nodes, and so on needed to support enterprise functions.

Transformation of work processes inherently must affect the operational view of the architecture. Changes of this view are likely to affect the technical and systems views. In contrast, changes of system and/or technical views that do not change operational views do not, by definition, change work processes. Hence, these types of changes may improve processes but do not transform the enterprise.

Changing the tasks and activities of the enterprise, by themselves, relates to business process improvement. In contrast, changing the purpose, objectives, and/or functions of the enterprise is more likely to be transformational. Such changes may, of course, cause tasks and activities to then change. Thus, change at any level in the hierarchy is likely to cause changes at lower levels.

Ultimately, one could liquidate the enterprise and redeploy its financial and perhaps physical assets in other ventures. However, it is difficult to characterize this as transformation. Thus, there is a point at which the change is sufficiently substantial to conclude that the enterprise has been eliminated rather than transformed.

Finally, it is useful to note that the multilevel architecture of population health in Chapter 7 provides a broader view of architecture in that it includes elements of the context in Figure 9.1. This broader view is often essential to addressing change while other elements of the context are changing as well. The changing payment system in healthcare provides an interesting illustration of providers pursuing transformation while the external context is changing, and not always predictably.

Management Decision-Making

Value deficiencies and work processes define the problem of enterprise transformation—one should recognize and/or anticipate deficiencies and then redesign work processes to remediate these deficiencies. To fully understand transformation, however, we need to understand both the problem and the problem-solvers. Thus, a characterization of management decision-making is central to our overall theory.

Mintzberg's (1975) classic paper, as well as more recent works (Mintzberg, Ahlstrand, & Lampel, 1998; Mintzberg & Lampel, 1999), serves to shatter the myth of the manager as a coolly analytical strategist, completely focused on optimizing shareholder value and using leading-edge methods and tools. Simon (1957, 1969) articulates the concept of "satisficing," whereby managers find solutions that are "good enough" rather than optimal. Another important factor is the organizational environment, which can be rife with delusions that undermine strategic thinking (Rouse, 1998).

This somewhat skeptical view of management decision-making ignores several important aspects of human decision-making. Managers' expertise and intuitions (Klein, 2002) and abilities to respond effectively in a blink (Gladwell, 2005) can be key to success, especially in recognizing what is really happening in an enterprise. Managers' roles as leaders, rather than problem-solvers and decision-makers, are also central to transformation (Kouzes & Posner, 1987).

Summarizing, the problem of transformation (i.e., value deficiencies prompting the redesign of processes) combines with the nature of the problem-solvers addressing transformation, as well as their organizations, to determine whether transformation is addressed, how it is addressed, and how well desired outcomes are achieved. The key point is that explanations of any particular instance of transformation will depend on the situation faced by the enterprise, the nature of the particular managers leading the enterprise, and the social structure of the enterprise.

Summary of Theory

Figure 9.3 summarizes the theory of transformation outlined in this chapter. Transformation is driven by value deficiencies and involves examining and changing work processes. This examination involves consideration of how changes are likely to affect future states of the enterprise. Potential impacts

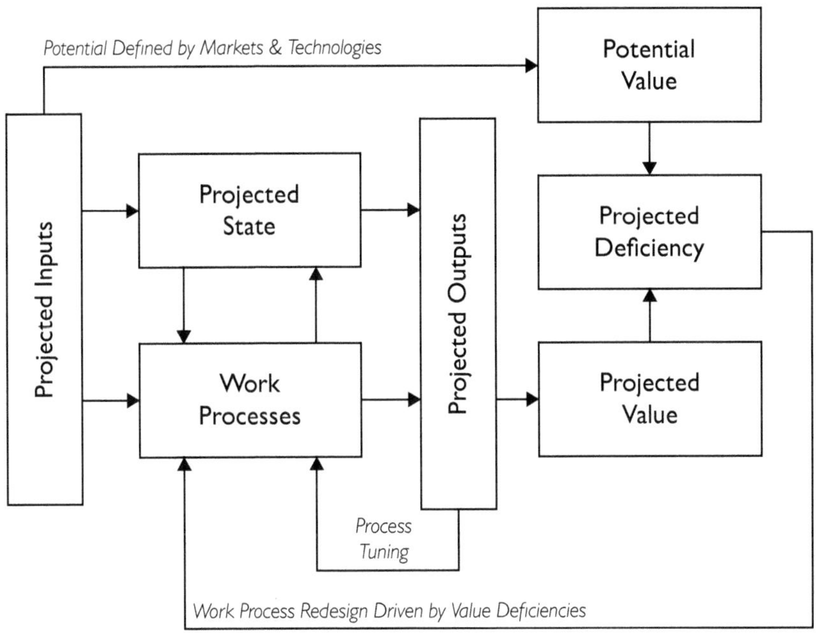

Fig 9.3 Theory of enterprise transformation.

on enterprise states are assessed in terms of value consequences. Projected consequences can, and should, influence how investments of attention and resources are allocated. The problem-solving and decision-making abilities of management, as well as the social context, influence how and how well all of this happens.

Ends, Means, and Scope of Transformation

There is a wide range of ways to pursue transformation. Figure 9.4 summarizes conclusions drawn from numerous case studies. The ends of transformation can range from greater cost efficiencies, to enhanced market perceptions, to new product and service offerings, to fundamental changes of markets. The means can range from upgrading people's skills, to redesigning business practices, to significant infusions of technology, to fundamental changes of strategy. The scope of transformation can range from work activities, to business functions, to overall organizations, to the enterprise as a whole.

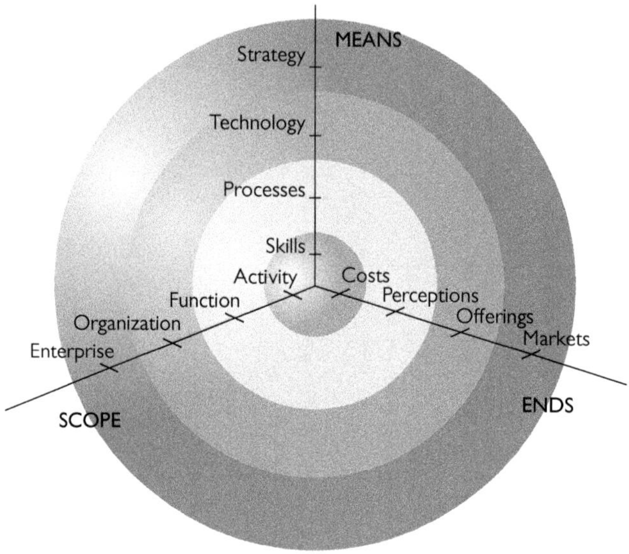

Fig 9.4 Transformation framework.

The framework in Figure 9.4 has provided a useful categorization of a broad range of case studies of enterprise transformation. Considering transformation of markets, Amazon leveraged IT to redefine book buying, while Walmart leveraged IT to redefine the retail industry. In these two instances at least, it can be argued that Amazon and Walmart just grew; they did not transform. Nevertheless, their markets were transformed. The U.S. Department of Defense's effort to move to capabilities-based acquisition (e.g., buying airlift rather than airplanes) has the potential to transform both the department and its suppliers.

Illustrations of transformation of offerings include UPS moving from being a package-delivery company to a global supply chain management provider; IBM's transition from manufacturing to services; Motorola moving from battery eliminators, to radios, to cell phones; and CNN redefining news delivery. Examples of transformation of perceptions include Dell repositioning computer buying, Starbucks repositioning coffee purchases, and Victoria's Secret repositioning lingerie buying. The many instances of transforming business operations include Lockheed Martin merging three aircraft companies, Newell Rubbermaid resuscitating numerous home products companies, and Interface adopting green business practices.

The costs and risks of transformation increase as the endeavor moves farther from the center in Figure 9.4. Initiatives focused on the center will typically involve well-known and mature methods and tools from industrial engineering and operations management. In contrast, initiatives toward the perimeter will often require substantial changes of products, services, channels, and so on, as well as associated large investments.

It is important to note that successful transformations in the outer band of Figure 9.4 are likely to require significant investments in the inner bands also. In general, any level of transformation requires consideration of all subordinate levels. Thus, for example, successfully changing the market's perceptions of an enterprise's offerings is likely to also require enhanced operational excellence to underpin the new image being sought. As another illustration, significant changes of strategies often require new processes for decision-making, for example, for R & D investments.

Value Deficiencies Drive Transformation

Elaborating earlier value-centered arguments, there are basically four alternative perspectives that tend to drive needs for transformation:

- value opportunities: the lure of greater success via market and/or technology opportunities prompts transformation initiatives;
- value threats: the danger of anticipated failure due to market and/or technology threats prompts transformation initiatives;
- value competition: other players' transformation initiatives prompt recognition that transformation is necessary to continued success; and
- value crises: steadily declining market performance, cash flow problems, and so on prompt recognition that transformation is necessary to survive.

The perspectives driven by external opportunities and threats often allow pursuing transformation long before it is forced on management, increasing the chances of having resources to invest in these pursuits, leveraging internal strengths, and mitigating internal weaknesses. In contrast, the perspectives driven by external competitors' initiatives and internally caused crises typically lead to the need for transformation being recognized much later and, consequently, are often forced on management by corporate parents, equity markets, or other investors. Such reactive perspectives on transformation often lead to failures.

Work Processes Enable Transformation

Transformation initiatives driven by external opportunities and threats tend to adopt strategy-oriented approaches, such as

- markets targeted, for example, pursuing global markets such as emerging markets, or pursuing vertical markets such as aerospace and defense;
- market channels employed, for example, adding web-based sales of products and services such as automobiles, consumer electronics, and computers;
- value proposition, for example, moving from selling unbundled products and services to providing integrated solutions for information technology management; and
- offerings provided, for example, changing the products and services provided, perhaps by private labeling of outsourced products and focusing on support services.

On the other hand, transformation initiatives driven by competitors' initiatives and internal crises tend to adopt operations-oriented approaches, including

- supply chain restructuring, for example, simplifying supply chains, negotiating just-in-time relationships, and developing collaborative information systems;
- outsourcing and offshoring, for example, contracting out manufacturing and information technology support, and employing low-wage, high-skill labor from other countries;
- process standardization, for example, enterprise-wide standardization of processes for product and process development, R & D, finance, personnel, and so on;
- process re-engineering, for example, the identification, design, and deployment of value-driven processes, and the identification and elimination of non-value-creating activities;
- the use of web-enabled processes, for example, online, self-support systems for customer relationship management, inventory management, and so on.

It is essential to note, however, that no significant transformation initiative can rely solely on either of these sets of approaches. Strategy-oriented initiatives must eventually pay serious attention to operations. Similarly, operations-oriented initiatives must at least validate existing strategies or run the risk of becoming very good at something they should not be doing at all.

The above approaches drive reconsideration of work processes. Processes are replaced or redesigned to align with strategy choices. Operational approaches enhance the effectiveness and efficiency of processes. Of course, the possibilities of changing work processes depend greatly on the internal context of transformation. Leadership is the key, but rewards and recognition, competencies, and so on also have strong impacts on success. Social networks enormously affect implementation of change.

Work processes can be enhanced (by acceleration, task improvement, and output improvement), streamlined (by elimination of tasks), eliminated (by outsourcing), and invented (by creation of new processes). An example of acceleration is the use of workflow technology to automate information flow between process steps or tasks. An illustration of task improvement is the use of decision-aiding technology to improve human performance on a given process task (e.g., enabling the consideration of more options). Output improvement might involve, for example, decreasing process variability. Streamlining could involve transferring tasks to others (e.g., transferring customer service queries to other customers who have addressed similar questions). Elimination involves curtailing processes, for example, Amazon created online bookstores, thus eliminating the need for bookstore-related processes in their business. Invention involves creating new processes, for example, Dell created innovative build-to-order processes.

Computational Theory

The qualitative theory just elaborated provides a well-founded explanation of enterprise transformation, as well as a framework for organizing case studies and best practices. However, it does not enable quantitative predictions of the outcomes on transformation and, in particular, the impacts of alternative strategy choices.

These choices include predicting better, learning faster, and acting faster. Predicting better decreases uncertainties about future states of market demands and competitors' offering. Learning faster implies that knowledge gained is more quickly incorporated into enterprise competencies. Acting faster quickly turns predictions and knowledge into practice. These three strategy choices

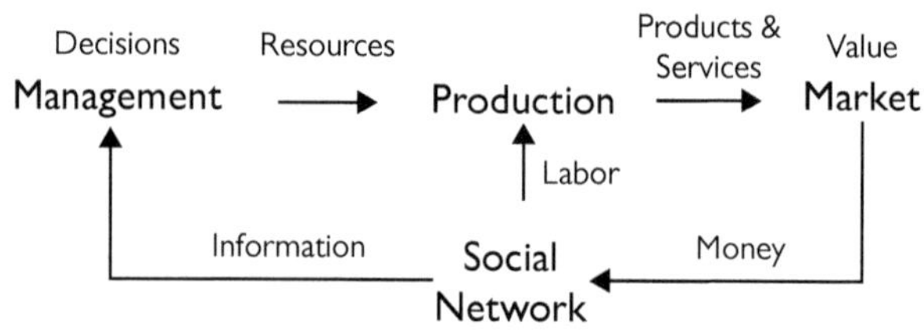

Fig 9.5 Elements of computational theory.

each require investments. We would like to know the conditions under which these investments make sense.

The key elements of the computational theory include the following:

- management: represents decisions to allocate resources to remediate value deficiencies and maximize expected utility;
- production: represents mapping of resources, including labor, to products and services over time;
- market: represents mapping from products and services to value that leads to revenues, profits, and cash flows over time; and
- social network: represents allocation of human attention to deploy resources, including provision of information for decision-making.

Figure 9.5 portrays the relationships among these elements. These elements were translated into a set of equations to enable a computational theory of enterprise transformation (Yu, Rouse, & Serban, 2011). The formulation of this set of equations drew upon several representational paradigms discussed in Chapter 2—dynamic systems theory, estimation theory, network theory, and decision theory. Thus, the enterprise and the marketplace are represented by a set of equations.

The computational theory elaborated predicts that companies will transform their enterprise by some combination of predicting better, learning faster, and acting faster, as long as the market is sufficiently predictable to reasonably expect that transformation will improve the market value the company can provide. If this expectation is unreasonable, then companies will sit tight and preserve resources until the market becomes more fathomable.

Put simply, if the market is highly volatile and customers are very discriminating, then the impacts of investments in transformation are likely to be negative, because it so difficult to hit a moving target in a turbulent environment. Companies sitting tight and preserving resources is often observed when the economy is weak, the market is hypercompetitive, and/or government policy is uncertain. In such situations, the computational model may show that the optimal strategy is to do nothing.

The computational theory is premised on the notion that companies make transformation decisions in response to the dynamic situations in which they find themselves. These decisions are affected by both what the company knows (or perceives) and the company's abilities to predict, learn, and act. Indeed, decisions to transform abilities to predict, learn, and act reflect desires to fundamentally change the company's overall ability to create market value. In this way, transformation decisions can enhance a company's abilities to address the ongoing and anticipated fundamental changes needed for success in dynamic markets.

This does not imply that everything will change. Instead, it means that everything needs to be considered in terms of how things consistently fit together, function smoothly, and provide high-value outcomes. This may be daunting, but it is entirely feasible. The key point is that one may not be able to transform value propositions without considering how the delivery enterprise itself should be transformed.

I hasten to note that, at this point, I am only addressing what is likely to have to change, not how the changes can be accomplished. In particular, the success of transformation initiatives depends on gaining the support of stakeholders, managing their perceptions and expectations, and sustaining fundamental change (Rouse, 2001, 2006, 2007). Leading initiatives where these factors play major roles require competencies in vision, leadership, strategy, planning, culture, collaboration, and teamwork (Rouse, 2011).

Conclusions

Fundamental transformation of a large enterprise is very difficult. Data on the Fortune 500 reported in The Economist supports this assertion (Schumpeter, 2009):

- In 1956–81, an average of 24 firms dropped out of the Fortune 500 list every year. This amounts to a 120 % turnover in that 25-year period.
- In 1982–2006, an average of 40 firms dropped out of the Fortune 500 list every year. This amounts to a 200 % turnover in the more recent 25-year period.

Thus, successful enterprise transformation is not only very difficult; it is becoming more difficult, and the failure rate is very high.

The extent of this difficulty is far from new. In Start Where You Are (Rouse, 1996), I summarize the changes addressed by roughly 200 enterprises over two centuries. Many of these enterprises tried to avoid being victims of "creative destruction" (Schumpeter, 1942), but almost all of them eventually failed. More contemporary case stories, with similar results, are discussed in Enterprise Transformation (Rouse, 2006).

Most enterprises eventually fail, but not every one fails. I discuss four detailed case studies of success at Lockheed Martin, Newell Rubbermaid, Reebok, and UPS in terms of drivers of transformation, approach to transformation, and elements of transformation (Rouse, 2011). The key elements of transformation in these four success stories were a focus on the customer, emphasis on operational efficiency, addressing the enterprise culture, and results-driven execution.

At the beginning of this chapter, I used Kodak, Polaroid, Xerox, Motorola, and Nokia as examples of companies that did not change and did not remediate their emerging value deficiencies. All of these companies had periods of great success, when revenues, profits, and share prices were soaring. Then, they were overtaken by creative destruction.

From a broad perspective, creative destruction is a powerful, positive force. New value propositions, often enabled by new technologies and led by new, innovative competitors, take markets away from established players. Jobs are created. The economy grows. People can, therefore, afford cars, TVs, smartphones, and so on.

The story is not as positive for the incumbents. They are under constant pressure. They have to face the dilemma of running the company they have while they try to become the company they want. But the company they have is usually consuming all the money and talent. As discussed in Chapter 5, they need to address the balance between investing in getting better at what they are already doing versus investing in doing new things.

It is very difficult to achieve this balance. Most of the stakeholders are strongly committed to the status quo. They need resources and attention to keep the status quo functioning. Many of the stakeholders in the future have yet to arrive. Consequently, they are not very demanding. Creating a sense of urgency is usually essential to addressing this stalemate.

Various pundits express this in the sense of needing a "burning platform." Rather than employing physical danger or the risks of imminent demise, one can use computational models to explore possible futures, both those that are desirable and those to be avoided. This can enable identifying leading indicators of both positive and negative changes. The result can be stories of change that, hopefully, everyone can understand and find compelling (Yu, Serban, & Rouse, 2013).

Key Points

- It is very difficult to successfully innovate when new offerings obsolete your current offerings.
- Enterprise transformation is driven by experienced and/or anticipated value deficiencies.
- The ends, means, and scope of transformation influence the difficulties, costs, and time required for fundamental change.
- Overall strategy choices include predicting better, learning faster, and acting faster; returns on investments in these areas depend on market conditions.
- Successful transformation initiatives usually require competencies in vision, leadership, strategy, planning, culture, collaboration, and teamwork.

REFERENCES

Drucker, P. F. (2001). *The essential Drucker: In one volume the best of sixty years of Peter Drucker's essential writing on management*. New York: Harper Business.

Gladwell, M. (2005). *Blink: The power of thinking without thinking*. Boston: Little, Brown.

Hammer, M., & Champy, J. (1993). *Reengineering the corporation: A manifesto for business revolution*. New York: Harper Business.

Hauser, J. R., & Clausing, D. (1988, May–June). The House of Quality. *Harvard Business Review*, 63–73.

Kaplan, R. S., & Norton, D. P. (1996, Jan–Feb). Using the Balanced Scorecard as a strategic management tool. *Harvard Business Review*, 75–85.

Klein, G. (2002). *Intuition at work: Why developing your gut instincts will make you better at what you do*. New York: Currency.

Kouzes, J. M., & Posner, B. Z. (1987). *The leadership challenge: How to get extraordinary things done in organizations*. San Francisco: Jossey-Bass.

Mintzberg, H. (1975, July–August). The manager's job: Folklore and fact. *Harvard Business Review*, 49–61.

Mintzberg, H. & Lampel, J. (1999, Spring). Reflecting on the strategy process. *Sloan Management Review*, 21–30.

Mintzberg, H., Ahlstrand, B., & Lampel, J. (1998). *Strategy safari: A guided tour through the wilds of strategic management*. New York: Free Press.
Rouse, W. B. (2005). A theory of enterprise transformation. *Journal of Systems Engineering, 8*(4), 279–295.
Rouse, W. B. (1998). *Don't jump to solutions: Thirteen delusions that undermine strategic thinking*. San Francisco: Jossey-Bass.
Rouse, W. B. (Ed.). (2006). *Enterprise transformation: Understanding and enabling fundamental change*. Hoboken, NJ, Wiley.
Rouse, W. B. (2001). *Essential challenges of strategic management*. New York: Wiley.
Rouse, W. B. (2011). Necessary competencies for transforming an enterprise. *Journal of Enterprise Transformation, 1*(1), 71–92.
Rouse, W. B. (2007). *People and organizations: Explorations of human-centered design*. New York: Wiley.
Rouse, W. B. (1996). *Start where you are: Matching your strategy to your marketplace*. San Francisco: Jossey-Bass.
Schumpeter, J. A. (1942). *Capitalism, Socialism and Democracy*. New York: Harper.
Schumpeter, J. A. (2009, September 19). Taking flight. *The Economist*, 78.
Simon, H. A. (1957). *Models of man: Social and rational*. New York: Wiley.
Simon, H. A. (1969). *The sciences of the artificial*. Cambridge, MA: MIT Press.
Slywotsky, A. J. (1996). *Value migration: How to think several moves ahead of the competition*. Boston: Harvard Business School Press.
Slywotsky, A. J., & Morrison, D. J. (1997). *The profit zone: How strategic business design will lead you to tomorrow's profits*. New York: Times Books.
Womack, J. P., & Jones, D. T. (1996). *Lean thinking: Banish waste and create wealth in your corporation*. New York: Simon & Schuster.
Yu, X., Rouse, W. B., & Serban, N. (2011). A computational theory of enterprise transformation. *Systems Engineering, 14*(4), 441–454.
Yu, X., Serban, N., & Rouse, W. B. (2013). The demographics of change: Enterprise characteristics and behaviors that influence enterprise transformation. *Journal of Enterprise Transformation, 3*(4), 285–306.

CHAPTER 10

Exploring Possible Futures

In this chapter, I discuss the nature of exploring possible futures, not in terms of case studies, but with a higher-level view of the overall process. I start by summarizing where we have journeyed in the earlier chapters. I then elaborate the nature of exploration, with emphasis on problem-solving, the impacts of technology advances, and the risks of exploration. I conclude with some suggested further readings.

Summary

In Chapters 3–9, I discussed explorations of possible futures in several domains:

- costs of higher education (Chapter 3);
- market success of products and services (Chapter 4);
- value of adopting new technologies (Chapter 5);
- impacts of failures of engineered systems; focused on training (Chapter 6);
- human health and well-being (Chapter 7);
- augmenting human intelligence; focused on aiding (Chapter 8); and
- transforming enterprise value propositions (Chapter 9).

The case studies in these chapters relied upon one or more of the eight modeling paradigms discussed in Chapter 2, although my focus was less on these paradigms than on the framing of the questions and formulations of the models associated with each application.

Computing Possible Futures. William B. Rouse. Oxford University Press (2019). © William B. Rouse 2019.
DOI: 10.1093/oso/9780198846420.001.0001

Table 10.1 ***Themes associated with key points in each chapter***

Theme of Key Point	Number of Key Points Incorporating the Theme	Chapter								
		1	2	3	4	5	6	7	8	9
Investments	8			X		X				X
Stakeholders	7		X		X			X		
Change	5			X						X
Decisions	5	X	X			X	X			
Design	5	X			X			X	X	
Prediction	5	X	X	X				X		
Phenomena	3		X							
Training and aiding	3						X		X	
Failure	2						X			
Innovation	2								X	X

Key points were summarized at the end of each chapter. There were 45 key points in all. Looking across these points, 10 themes emerge; these are listed in Table 10.1.

Six themes—investments, stakeholders, change, decisions, design, and prediction—are reflected in 35 of the key points. Clearly, exploring possible futures, at least as addressed in this book, concerns investments in pursing these futures involving multiple stakeholders and significant change. Models are employed to frame decisions, assess alternative designs, and make predictions about what *might* happen.

Exploration

I like to work with executives and senior managers who have exploration mindsets. They very much want to know what might be possible and how they might achieve it. They want to understand how others are exploring, what futures they are entertaining, and how their investments might affect the possibilities that I am exploring with them.

In Chapter 2, I argued that you need to balance exploiting current capabilities and assets versus exploring new capabilities and assets. One of key points in Chapter 5 is "You need to address the balance between investing in getting better at what you are already doing versus investing in doing new things." People with exploration mindsets keenly understand this point.

I have been involved with several types of explorations many times. One type concerns planning new generations of existing product lines, hopefully taking the central value proposition to new levels. A second type involves considering new application domains for existing technology capabilities, with a goal of surprising incumbents in these domains with high value offerings they had not thought possible.

A third type of exploration concerns assessing technology options for enabling new offerings. Often, such explorations address markets that do not yet exist but where the sentiment is that they will emerge. There is, however, uncertainty about what technologies will predominate.

Yet another type of exploration involves addressing major enterprises challenges. The higher education cost bubble is a good example. Training and aiding to support detection, diagnosis, and compensation for system failures is another example. And, of course, enterprise transformation is a general instance.

Desirability versus Feasibility

I almost always try to launch these explorations by focusing on desirability in terms of the "preference space" of major stakeholders. At the same time, I attempt to avoid consideration of feasibility in the sense of what is achievable in the "physical space" of the domain of interest. Thus, for example, I like to begin by examining what customers really want, rather than initially worrying about whether it is possible.

My philosophy is that you will not achieve beyond your vision. In fact, you might fall short of your vision. Hence, your vision should be ambitious. I have approached this with a simple construct.

- Start with the vision, Level C, an ambitious idea or set of ideas for what might be possible; it is likely very risky but is certainly highly desirable.
- Next, consider the baseline, Level A, a technology, product, service, or organization that you know you can deliver, where all risks are manageable.

- Then plan the bridging concept, Level B, which likely has technology, financial, and time risks but is a significant step from Level A toward Level C.

The overarching principle is to market Level C, sell Level B, and deliver Level A. This may sound devious, but it is not if you explicitly explain the principle to customers, perhaps saying, "We both want the vision, but we both know it is not yet possible. We will deliver Level A within a few months,[1] so that you will have some immediate capabilities. In parallel, we will be researching Level B to decrease risks. You will have Level B in a couple of years."

Another way to explain this idea is to architect Level C, develop Level A within a clear subset of this architecture, and research Level B so that a larger subset of this architecture is enabled. The preference space drives Level C, while the physical space constrains Level A. Apple provides a good example of having successfully executed this approach. Each generation of digital devices keeps us willingly locked into their offerings.

When Explorations Fail

Successful explorations have some subtleties. Some lead to market innovations in terms of products, services, and organizations. Others lead to well-informed plans for new generations of current products and services. Many lead to plans for which resources must be sought to enable execution.

Occasionally, explorations lead to recognitions that the idea that motivated the effort in a bad idea. I have found that a notable strength of model-based explorations is getting rid of bad ideas quickly. Much time can be wasted on bad ideas as the people around the table argue about the pros and cons of these ideas. Using models for exploration moves the arguments from across the table to up on the screen.

One of my long-time clients was a large electronics and communications company. We pursued an exploration of their entering the medical electronics market. Several model-based workshops were held over the course of a month using a couple of the tools in the Advisor Series. In the end, it was decided that it did not make sense to enter this market.

As we were wrapping up, I stopped by the office of the vice president who had led this effort. I asked him if he was disappointed in the outcome. He said that he

[1] The time periods noted here are, of course, dependent on the domain you are exploring, for example, new digital devices happen much faster than new airplanes.

found the whole process quite successful. "We teed up an idea we have had for years, worked it diligently, and proved to ourselves it was a bad idea, all within one month. That is a huge success. We don't need to talk about that anymore."

Observations on Problem-Solving

Let's consider the many case studies discussed in this book in terms of problems addressed and observations on the problem-solving of the teams involved. I can cluster my observations into starting assumptions, framing problems, and implementing solutions.

Starting Assumptions

Who are the stakeholders in the problem of interest and its solution? It is essential that one identify key stakeholders and their interests. All critical stakeholders need to be aligned in the sense that impacts on their interests are understood. Chapter 4 discussed an automotive example where a significant stakeholder was forgotten. One risks the possibility of substantial and perhaps unexpected resistance to change from stakeholders whose concerns are ignored.

Look at problems and solutions from the perspectives of stakeholders. How are they likely to be thinking? In the licensing of technologies discussed in Chapter 5, I have found it to be crucial to understand buyers' exercise costs. This suggests that the licensor might provide the licensee consulting services to exercise the option less expensively.

Articulate and validate assumptions. Significant risks can result when there are unrecognized assumptions. Many of the case studies involved validating assumptions before deciding to invest in development. This can sometimes be difficult when key stakeholders "know" what is best.

Understand how other stakeholders may act. The effectiveness of a strategy is strongly affected by competitors. This is well illustrated in several of the case studies. Having one or more team members play competitors' roles can often facilitate this.

Framing Problems

Define value carefully. Translating invention to innovation requires clear value propositions, as illustrated in the discussion of enterprise transformation.

Value needs to be framed from the perspective of the marketplace, not the inventors. Markets do not see their main role as providing money to keep inventors happy.

Think in terms of both the current business and possible future businesses. Current success provides options for future success, but perhaps with different configurations for different markets. Several of the case studies in Chapter 5 illustrated how current products and customers provide options for new products and customers.

Consider possibilities for customizing solutions for different customers and constituencies. Population health, as discussed in Chapter 7, required stratification and tailoring of processes to varying health needs. This is critical to the viability of population health offerings. Henry Ford, almost 100 years ago, was the last person to believe that everyone wanted exactly the same automobile.

Access and integrate available data sets on customers, competitors, technologies, and so on. Many of the case studies in Chapter 7 involved significant data integration, which often substantially increases confidence. Great insights can be gained by mining available data sets, including internal sets, publicly available sets, and purchasable sets.

Plans should include strategies for dealing with legacies. The status quo can be an enormous constraint because it is known, paid for, and in place. The discussions in Chapter 9 illustrated the need to get legacies "off the books." Discarding or liquidating assets for which one paid dearly can be painful.

Implementing Solutions

It can be great fun to pursue market and/or technology opportunities. Innovators can earn high payoffs, albeit with high risks, as depicted in Chapter 5. The key is to have the human and financial resources to support and sustain the commitment to innovate.

In stark comparison, crises are not fun. The discussion of enterprise transformation in Chapter 9 illustrated the high costs and substantial consequences of delaying change. Often, the status quo has devoured most available human and financial resources. When change is under-resourced, failure is quite common.

The existing enterprise can hold change back. Chapter 9 portrayed the difficulty of changing business models. New business opportunities may be very attractive, but if success requires substantially new business models, one should assess the enterprise's abilities to make the required changes. Several of my clients avoided pursuing changes because these changes would have required substantial cultural changes.

Change should involve stopping as well as starting things. Stopping things will likely disappoint one or more stakeholders. The academic strategies case study in Chapter 3 illustrated the difficulty of keeping everybody supportive. The consequence is that the status quo dominates, especially when senior management team members were recruited to be stewards of the status quo.

Impacts of Technology Advances

In 2014, I gave a presentation at a colloquium dedicated to Norbert Wiener, in honor of his birth 120 years earlier. This caused me to revisit Wiener's pioneering contributions to control theory, as well as Claude Shannon's formulation of information theory, Philip Morse's work in queuing theory, and other researchers' roles in advancing many of the paradigms discussed in Chapter 2.

I was struck by how many of our contemporary ideas and theories were being discussed in the 1940s and 1950s. Our ideas are not that new and unique. What are new are the enormous computing power available, pervasive big data and analytics, and amazing visualization technology. Wiener and Shannon had to make due with pencil and paper, perhaps augmented with mechanical computing devices that they designed and built themselves.

The layered statistical models underlying machine learning are cutting edge, but the idea dates to the 1950s and 1960s. Algorithms and other software techniques have advanced, while computer and communications technologies have become pervasive. These trends will likely continue. For example, quantum computing will provide more power but will not inherently solve the challenges addressed in this book.

I think that evidence-based and model-based decision-making will become increasingly pervasive. An increasing portion of decision-makers will expect to see data visualizations and be able to manipulate them to explore possible futures. There is a risk that a decreasing proportion of the general population will understand such presentations. Competitiveness will drive the former, but education will be needed to address the latter.

Risks of Exploration

In Chapter 9, I discussed the promises, perils, and prospects for AI and machine learning, as well as analytics in general. The promise of pervasive data should be balanced with the perils of modeling traps, bad data, and misuse of data. Socially responsible automation should be the goal. This section considers risks more broadly in terms of model creation, use, and maintenance.

Model Creation

Models can be flawed in several ways. The computations of solutions can be faulty. The data used to parameterize models can be biased or outright wrong, as discussed in Chapter 8. Model-induced design errors can be due to ideas that work in the model world but not in the real world.

Phenomena can be over-modeled in the sense that much more detail is included than is useful, bogging down computations and limiting experimentation. When we conducted the workshop that led to the book *Organizational Simulation* (see "Further Reading"), one participant was a Hollywood producer. He commented on the penchant to model every blade of grass despite the fact that grass is not central to the idea being pursued.

Over-modeling often results when the question of interest is not clear. One visitor to my office explained how each run of his model took one week. I asked why. He explained many of the details being simulated. I asked what question motivated him to include these details. He responded, "I was told to develop the model. I was not told what it would be used for."

Model Use

It is extremely rare for any exploration of the future to not involve uncertainty. Yet, many modelers only predict expected values, that is, means of future system states. This can easily result in misplaced confidence in the results. Ideally, you should compute probability distributions of future states. This complicates the modeling effort, but it is usually important.

Pursuing optimization when it is unwarranted can yield projections that are more crisp than they are defensible. Decision-makers sometimes like point estimates, that is, a single number rather than a distribution. I recall a CFO asking me what the likelihood was that the return on investment would be exactly the mean of the distribution we projected. I said, "Zero!" He looked shocked. Then, I told him that the probability of any single point on an infinite, continuous scale is zero. We can only calculate probabilities of a variable occurring in a range.

There are also risks of overreliance on models. This tends to occur when models have been repeatedly employed over extended periods of time. Users tend to develop intuitions about the model world rather than the real world, limiting their value in providing "sanity tests" of model projections. Fortunately, blind acceptance of model projections is quite rare.

Model Maintenance

Maintaining the currency of models tends to be challenging. Ongoing effort is needed to update estimates of model parameters based on more recent data or on data from a broader set of sources. Models also tend to have fading provenance. Key assumptions become obscure, and documentation gets out of date. Once the model developer is out of the picture, organizational memory fades as well.

There tends to be overreliance on legacies, in terms of both modeling tools and specific instantiations of models. This is due, in part, to the high costs of switching from one tool to another, and especially from one paradigm to another. For example, switching from a dynamic system theory representation to a decision theory representation would tend to be a major undertaking, which most clients would like to avoid.

Uncharted Territory

An existential risk is that you uncover inconvenient truths. For instance, you learn that changes in your markets will be much more disruptive than anticipated, as was discussed in Chapter 3 for higher education. You might find, for example, that your technology does not provide the competitive advantages you expected, in part because your competitors' strategies do not play to your strong suit. I discussed a couple examples of this in Chapter 4.

This can result in unhappy clients. They were interested in possible futures that you have shown to be very unlikely, perhaps even impossible. I recall one instance where the decision-maker provided a rather negative assessment of my findings. I asked, "Do you doubt my data and computations?" He responded, "No, I believe them. I just don't like them."

This is a good time to be creative. In one of my engagements, we were developing a plan to invest in technology that would circumvent a competitor's patents. Our projections indicated that this investment would be risky and costly. The vice president leading this effort was getting frustrated. I asked about the size of the competitor. They were a much smaller company and not doing very well.

I suggested that one option be buying the competitor. Our analyses showed that this was a much better investment than the technology workaround strategy. My client subsequently acquired the smaller company. The R & D planning meeting had been transformed into an M & A meeting. This was uncharted territory for me.

I learned a particularly important lesson from this engagement. It is very useful to have a deep enough understanding of the representations and data underlying your models to be able to assess their utility in addressing questions for which these models were not developed. The tool we were using was not developed for M & A analyses, but I saw how I could configure the input data (spreadsheet projections) to get the model to yield the right answers.

Conclusions

We all use models to address questions such as "What will happen if I pursue a certain course of action?" Many of the models we employ are "mental models." For example, our mental models enable us to readily ride a bicycle despite not having ridden one in quite some time. We have mental models for driving cars, using a computer, and many other tasks. We also have mental models that help us anticipate situations, for instance, the likely course of a business meeting.

When the effort is warranted, we externalize our models, formally represent them, create computational forms, and explore alternative answers to our questions. This book has elaborated this process. My case studies provided readers with a sense of how computational models can be used, for instance, to explore transformation of the U.S. healthcare system and the likely impacts of driverless cars on the U.S. economy.

Underpinning all the examples and case studies have been in-depth reflections on what you are doing when you develop models, how you can only predict what *might* happen—possible futures—and how computational models and their predictions can best inform human decision-making. For example, models can help you to understand leading indicators of particular futures, or tipping points that make some futures more likely than others.

Model-based problem-solving and decision-making provide powerful means to explore possible futures. They require that assumptions be explicit and data be vetted. They enable sharing and manipulating visualizations, getting rid of bad ideas quickly and refining good ideas for subsequent empirical validation. They support teamwork, particularly by enhancing team mental models. Finally, they enable crossing disciplinary boundaries. Such border crossings are often the hallmarks of innovation.

FURTHER READING

I have focused on modeling concepts and case studies, and avoided mathematics, in this book. There is a wealth of books with considerable mathematical and algorithmic content. The following sources represent a middle ground with more technical detail than this book, but limited mathematical details. Of particular importance, several of these sources discuss deeper theoretical and philosophical issues than those addressed in this book. I recommend these sources to you.

Alpaydin, E. (2016). *Machine learning*. Cambridge, MA: MIT Press.

Davis, P. K., O'Mahony, A., & Pfautz, J. (Eds.). (2019). *Social behavioral science modeling*. Hoboken, NJ: Wiley.

DMDU Society. (2019). *Society for Planning Under Deep Uncertainty*. Retrieved January 12, 2019, from http://www.deepuncertainty.org/.

Mittal, S., & Tolk, A. (Eds.). (2019). *Complexity challenges in cyber physical systems*. Hoboken, NJ: Wiley.

Mittal, S., Diallo, S., & Tolk, A. (Eds.). (2018). *Emergent behavior in complex systems engineering: A modeling and simulation approach*. Hoboken, NJ: Wiley.

RAND. (2019). *Robust decision making*. Santa Monica, CA: RAND Corporation. Retrieved January 12, 2019, from http://www.rand.org/topics/robust-decision-making.html.

Rouse, W. B. (2015). *Modeling and visualization of complex systems and enterprises: Explorations of physical, human, economic, and social phenomena*. Hoboken, NJ: Wiley.

Rouse, W. B., & Boff, K. R. (Eds.). (2005). *Organizational simulation: From modeling and simulation to games and entertainment*. New York: Wiley.

Sheridan, T. B. (2017). *Modeling human system interaction: Philosophical and methodological considerations, with examples*. Hoboken, NJ: Wiley.

Tolk, A. (Ed.). (2013). *Ontology, epistemology, and teleology for modeling and simulation*. Berlin, Heidelberg: Springer-Verlag.

Zeigler, B. P., Praehofer, H., & Kim, T. G. (2000). *Theory of modeling and simulation: Integrating discrete event and continuous complex dynamic systems*. New York: Academic Press.

INDEX